WESTWOOD

Project Management in the Fast Lane

Applying the Theory of Constraints

The St. Lucie Press/APICS Series on Constraints Management

Series Advisors

Dr. James F. Cox, III
University of Georgia
Athens, Georgia

Thomas B. McMullen, Jr.
McMullen Associates
Weston, Massachusetts

Titles in the Series

Introduction to the Theory of Constraints (TOC) Management System
by Thomas B. McMullen, Jr.

Securing the Future: Strategies for Exponential Growth Using the Theory of Constraints
by Gerald I. Kendall

Project Management in the Fast Lane: Applying the Theory of Constraints
by Robert C. Newbold

The Constraints Management Handbook
by James F. Cox, III and Michael S. Spencer

Thinking for a Change: Putting the TOC Thinking Processes to Use
by Lisa J. Scheinkopf

Management Dilemmas: The Theory of Constraints Approach to Problem Identification Solutions
by Eli Schragenheim

Project Management in the Fast Lane

Applying the Theory of Constraints

Foreword by
Thomas B. McMullen, Jr.

Robert C. Newbold

The St. Lucie Press/APICS Series on Constraints Management

S^t_L

St. Lucie Press
Boca Raton   Boston   London   New York   Washington, D.C.

Library of Congress Cataloging-in-Publication Data

Newbold, Robert C.
    Project management in the fast lane : applying the theory of
constraints management / by Robert C. Newbold.
        p. cm. — (St. Lucie Press/APICS series on constraints management)
    Includes bibliographical references and index.
    ISBN 1-57444-195-7 (alk. paper)
    1. Industrial project management. I. Title. II. Series.
HD69.P75N484   1988
658.4′04--dc21
                                       97-48806
                                         CIP

© 1998 by CRC Press LLC
St. Lucie Press is an imprint of CRC Press LLC

No claim to original U.S. Government works
International Standard Book Number 1-57444-195-7
Library of Congress Card Number 97-48806
Printed in the United States of America   3 4 5 6 7 8 9 0
Printed on acid-free paper

St Lucie Press
2000 Corporate Blvd., N.W.
Boca Raton, FL 33431-9868

APICS
500 West Annandale Road
Falls Church, Virginia 22046-4274

Dedication

To my family, including Jim, Tom, and Don.

Contents

SECTION IV: IMPLEMENTATION ISSUES

APPENDICES

Foreword

At Last!

At last! A book that documents what only a few companies know. Rob Newbold's new book gives project managers the practical and hands-on detail they need to put Dr. Eliyahu M. Goldratt's TOC critical chain project management and scheduling breakthroughs to work.

Management Science

The Theory of Constraints (TOC) is a new and important expression of management science invented by Dr. Eliyahu M. Goldratt, a scientist, physicist, author, educator, and consultant. Since the mid-1970s, Dr. Goldratt has used scientific methods to create concepts in management which have proven to be of great value to industry. He has encouraged and inspired others to use scientific thinking methods in their professional and personal lives. He has led and encouraged efforts to apply scientific thinking methods to areas outside the traditional "hard" sciences. These have so far included the disciplines of general management, manufacturing management, manufacturing information systems, management accounting, project management, day-to-day managing skills, administration and content of education at various levels, and several aspects of the relationship between thinking and health. These are areas in which the thinking methods of science have seldom been used, or have — at least, arguably — rarely been used effectively. Finally, Dr. Goldratt has invented his own expression of the scientific method, the structured TOC Thinking Processes (TP). These thinking processes take the form of the family of TOC "logic tree" management processes and diagrams.

These tools make the scientific method more understandable and practical. This makes the science approach more effective for day-to-day use, by many more people, in all walks of life, all over the world. Examples of non-industrial applications by individuals include a successful attempt to earn a place on an Olympic swim team, rapid improvements in little league baseball batting performance, dramatic improvements in conflict management in high schools, and a variety of breakthroughs in mental and physical health. Since the scope of application is any area of life affected by the way people think about it, the Theory of Constraints — and especially its structured thinking processes — may usefully be considered a physics of anything.

Critical Chain vs. Critical Path

This book describes the results of applying TOC to the domain of project management. The standard concepts in project scheduling and management have been CPM and PERT. The concept of critical chain removes, among other assumptions, the implicit assumption of infinite capacity from the project management domain, just as the well-known TOC drum-buffer-rope removed it from the factory management domain. Rob Newbold's new book in the St. Lucie Press/APICS Series on Constraints Management, *Project Management in the Fast Lane: Applying the Theory of Constraints,* is the best, most practical, and most detailed reference on this project.

Long-Awaited Support for Software Developers

This book also provides software development and marketing companies with the information they need to add TOC's critical chain project management modules to their line of TOC product offerings. Critical chain modules are excellent complements to existing suites of TOC software products in the logic tree thinking process, throughput value added (TVA) management accounting, constraints-based supply chain, drum-buffer-rope finite capacity scheduling, and buffer management operations control areas.

Rob Newbold knows of what he writes. He is one of the reasons software support for critical chain project scheduling is now starting to become available in the market. This book will accelerate this trend by making the technology clear to software companies all over the world.

Keeping Projects on Track

Mistakes in a large project's basic assumptions can be very expensive. It is far better to identify any needed adjustments at the very beginning of a project — through thinking alone — than to use the subsequent investment of thousands or millions of dollars, as the company's "trial and error" process for coming to the same conclusions. In this arena, even small improvements in the efficiency or quality of thinking, consensus, and communication processes are extremely valuable. It's awful when a project starts wrong, or dies too late. There are many examples of projects that died merciful deaths, but still after too much money and time had been invested in a wrong solution.

Real thinking really works!

> **Thomas B. McMullen, Jr.**
> *Vice President-Education Development (SIGs)*
> *APICS — The Educational Society for Resource Management*

Preface

I used to have a secret formula for estimating how long jobs would take: I always multiplied my first "reasonable" estimate by 2½. It worked. In order not to run out of things to do, I always had several projects going at once. That worked too. Both of these practices are common. I have been very successful using common practice to manage software development projects; many others have used it successfully as well, in many fields. So why write a book that shows, not just that I was wrong, but how wrong I was? Why try to improve the tried and true?

There are several reasons for this book. First, there are better ways to manage projects. These better ways have worked for me, they have worked for my clients, and they have worked for many others around the world. Second, the traditional ways of managing projects don't guarantee success; with the world changing more and more quickly, they will eventually guarantee failure. And third, these ideas are unfortunately not well known. I want to see them become common practice, as I think eventually they must.

This book is about Theory of Constraints (TOC). TOC is much more a management philosophy than a set of techniques. The philosophy is partly embodied in the "three fundamental questions": what to change, to what to change, and how to cause the change. This book is organized around those three questions.

Part I deals with the question "what to change?" Common current problems and behavior patters are examined, in order that we can better understand the underlying causal relationships and core problems.

Next we need solutions to the problems; we need to understand "to what to change." One of the most common problems is uncertainty. We must learn to deal with uncertainty in such a way that its effects aren't constantly thwarting

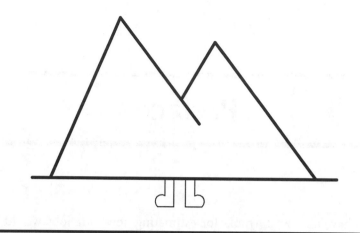

Figure 1   The Mountain Comes to Mohammed

our efforts to improve. Part II goes into some detail describing means of dealing with uncertainty in project scheduling and management.

The most important message of Part III is the importance of taking a global viewpoint. For the "to what to change" to make sense, for the actions to be relevant, those actions must relate to some global objectives: a goal. Part III describes goals and presents some processes which are useful to help manage an organization according to its goal.

We must also be able to implement the ideas. We must understand the desired consequences, the needed actions and the pitfalls. Significant changes won't — or certainly shouldn't — be attempted without that understanding. Unintended consequences can be very unpleasant, as illustrated in Figure 1.

While a detailed implementation plan will be specific to the individuals or organizations it applies to, many lessons have been learned and can be shared. Part IV deals with some common implementation issues, problems and ideas.

Many thanks and great credit go to Dr. Eliyahu M. Goldratt, to whose unflagging efforts and phenomenal insights the existence of Theory of Constraints must be attributed. Thanks also go to my indefatigable friends and colleagues at The TOC Center and the Avraham Y. Goldratt Institute, whose deep thinking and hard work have helped make Theory of Constraints ever more popular, powerful and generally applicable. While I can honestly say there are far too many people deserving of thanks to mention specifically, nevertheless specific thanks go to Bob Fox, Dale Houle, Dee Jacob and Dick Moore. They have all, at many times and in many ways, been generous with their time and ideas.

There are many associates of The TOC Center and the Goldratt Institute, past and present, whose work and hard-won experience have contributed immeasurably to the base of TOC knowledge that I am trying to share. People like Jim Cox, Tom McMullen, Tony Rizzo, Lisa Scheinkopf, Eli Schragenheim, Bob Stein and many, many others have been deeply involved in this process.

The publisher, St. Lucie Press, certainly deserves a great deal of credit, not only for their hard work but for taking a chance with this book. Heartfelt thanks also go to those who have looked at these materials in various stages and given their comments and support. This group includes Brendan Hemingway, Bill Lynch, Gabriel Lopez, Ken Pasterczyk and Bob Vornlocker.

Finally, praise and love goes to my family for their unflagging support over the inordinate amount of time it took to complete this project.

About the Author

Robert C. Newbold has over 20 years experience as a consultant, analyst, manager and software developer in the defense, health care and manufacturing fields. From 1988 to 1996 this expert worked as manager and developer with the Avraham Goldratt Institute and the TOC Center, creating and implementing scheduling and control tools for manufacturing management. In April of 1996 he started his own company, Creative Technology Labs, LLC, in order to develop improved software tools for project managers. Mr. Newbold holds degrees from Stanford, Yale and SUNY Stony Brook, and is a sought-after speaker.

ABOUT APICS

APICS, The Educational Society for Resource Management, is an international, not-for-profit organization offering a full range of programs and materials focusing on individual and organizational education, standards of excellence, and integrated resource management topics. These resources, developed under the direction of integrated resource management experts, are available at local, regional, and national levels. Since 1957, hundreds of thousands of professionals have relied on APICS as a source for educational products and services.

- APICS Certification Programs — APICS offers two internationally recognized certification programs, Certified in Production and Inventory Management (CPIM) and Certified in Integrated Resource Management (CIRM), known around the world as standards of professional competence in business and manufacturing.
- APICS Educational Materials Catalog — This catalog contains books, courseware, proceedings, reprints, training materials, and videos developed by industry experts and available to members at a discount.
- APICS — The Performance Advantage — This monthly, four-color magazine addresses the educational and resource management needs of manufacturing professionals.
- APICS Business Outlook Index — Designed to take economic analysis a step beyond current surveys, the index is a monthly manufacturing-based survey report based on confidential production, sales, and inventory data from APICS-related companies.
- Chapters — APICS' more than 270 chapters provide leadership, learning, and networking opportunities at the local level.

- Educational Opportunities — Held around the country, APICS' International Conference and Exhibition, workshops, and symposia offer you numerous opportunities to learn from your peers and management experts.
- Employment Referral Program — A cost-effective way to reach a targeted network of resource management professionals, this program pairs qualified job candidates with interested companies.
- SIGs — These member groups develop specialized educational programs and resources for seven specific industry and interest areas.
- Web Site — The APICS web site at http://www.apics.org enables you to explore the wide range of information available on APICS membership, certification, and educational offerings.
- Member Services — Members enjoy a dedicated inquiry service, insurance, a retirement plan, and more.

For more information on APICS programs, services, or membership, call APICS Customer Service at (800)444-2742 or (703)237-8344 or visit http://www.apics.org on the World Wide Web.

Introduction

The measures of success for a project manager are very simple: just complete your projects within budget and on time, while keeping the customers happy. Cost, schedule, and performance. The problems project managers run into while trying to meet these simple requirements are anything but simple, either to enumerate or to understand. In fact, the problems are widely regarded as so intractable and inevitable that they are treated as accepted parts of our reality. Is it "normal" in many environments for projects to be late? Are "crunch" times part of life? Is burnout an inevitable result of the pressures of today's competitive business environment?

While companies are usually reluctant to talk about their performance on projects, the auditing organizations of governments and public institutions are paid to generate this information. These sources seem to validate the inevitability of poor project results. For example,

> U.S. Federal Aviation Administration: "Since the 1980s, FAA's modernization efforts experienced substantial cost overruns and lengthy schedule delays. The cornerstone project in FAA's plan was the Advanced Automation System (AAS), a major program to replace all the hardware and software in FAA's tower, terminal, and en route facilities. AAS was initiated in the early 1980s. By the early 1990s the estimated cost for AAS increased from the original $2.5 billion estimated in 1983 to $7.6 billion. The program was approximately 8 years behind the original schedule."[1]
>
> Information technologies: "During the past 6 years, agencies have obligated over $145 billion building up and maintaining their information technology infrastructure. The benefits from this vast expenditure, however, have frequently been disappointing. GAO reports and congressional

hearings have chronicled numerous system development efforts that suffered from multimillion dollar cost overruns, schedule slippages measured in years, and dismal mission-related results."[2]

Defense projects: "Despite DOD's past and current efforts to reform the acquisition system, wasteful practices still add billions of dollars to defense acquisition costs. Many new weapon systems cost more and do less than anticipated, and schedules are often delayed."[3]

U.S. Department of Energy: "GAO found that: (1) from 1980 through 1996, DOE conducted 80 projects that it designated as major system acquisitions; (2) DOE has completed 15 of these projects, and most of them were finished behind schedule and with cost overruns; (3) 31 other projects were terminated prior to completion after expenditures of over $10 billion; (4) cost overruns and schedule slippages continue to occur on many of the ongoing projects; …"[4]

Projects funded by the World Bank: "Time overruns have gone down, but forecasts are still overoptimistic. On average, operations evaluated in 1994 took 37 percent longer to implement than originally scheduled, down from 48 percent in 1993, and 54 percent in the 1974-94 cohorts."[5]

The projects represented here form a wide cross section of project environments and objectives; there is no reason to suppose that late, over-budget, and under-performance projects are unusual. Everyone assumes their situation is different, their environment has valid reasons for being difficult to manage. For example, one report reasons that "large construction projects are generally prone to budget and schedule overruns. This may stem from the fact that construction projects are unlike the products of most manufacturing and industrial projects."[6] Everyone wonders why it happens, and everyone has an answer, sometimes many answers, sometimes a new answer each month. Typically the different problems are attributed to various causes. Frequently these separate causes are attacked with vigor through various actions:

Cause	Action
Vendor problems	Better contracts, more detailed supervision, more suppliers, sole-source suppliers
Scope changes	Better statement of work, better monitoring
Uncertainty	Risk analysis, more detailed specifications, more planning, less planning
Lack of teamwork	Focused teams, incentive plans
High operating expenses	Downsizing, more efficiency, moving offshore
Bad weather	Better luck next time

These actions are not "wrong." Indeed, under the right circumstances many of them produce real benefits. But we also have to accept that neither the problems nor the solutions are new. All of the solutions mentioned in the table above, in one form or another, have been around for at least 20 years. Since problems still exist, the solutions must be inadequate.

Given the increased global competition experienced today in virtually all business sectors, there is ever more pressure to produce more, more cheaply, more quickly. This increased pressure, when projects are already under performance, over budget, and over schedule is only likely to make problems worse as managers try ever harder to implement ineffective solutions. The prognosis looks disappointing.

It's easy — too easy — to believe that there can be no general answer, not only because the problems have been around for so long, but because of the wide range of settings in which they occur. The project field, and the problems with project management, are even more extensive than the quotes above would suggest. We all work on projects throughout our lives. Our projects could be anything from home renovations to a space shuttle launch to a book report. It is difficult to accept that solutions that can work in aerospace can work in construction or computer programming or planning a vacation. We seem to have innumerable problems and a consequent need for many solutions. And yet we have to wonder, because the problems in so many fields look so similar.

A basic principle of Theory of Constraints (TOC) is that the unpleasant problems or "undesirable effects" we experience in a field such as project management are usually the result of relatively few core problems. "Relatively few" means a manageable number. If we can identify these few core problems, and can address them, the majority of the undesirable effects will go away. By addressing the underlying problems, rather than the symptoms, we address many problems simultaneously. We have our magic bullet.

A simple example might help convey the idea. Suppose I have severe problems maintaining my yard. My grass won't grow, the leaves are a lot of trouble to rake, and worst of all the windows are always dirty because the birds seem to use them as restrooms. I could continue to rake the leaves, buy grass seed, and clean the windows. Or I could remove the tree that provides the leaves and the shade and the bird's homes, and solve all three problems.

TOC is a common-sense combination of techniques and philosophy that can enable dramatic, rapid, and ongoing improvements by helping identify these core problems and by providing tools that allow the development of workable solutions. TOC and its forerunner, Synchronous Manufacturing, have been around for close to 20 years. The first TOC concepts related to manufacturing and shop-floor scheduling, but over the years both specific applications and generic problem-solving techniques were developed.

During this time TOC has been used by more and more companies around the world to improve bottom-line results, primarily in the manufacturing arena. The largest companies, such as Ford, AT&T, and the U.S. military have used TOC extensively; very small companies and individuals have used it as well. They have good reason. It is not uncommon for the application of TOC to reduce factory inventory levels by 75% or more and increase throughput by 40%. These dramatic improvements frequently come after more conventional approaches have failed.

Success breeds acceptance. Since the early days in the late 1970s, many of the "revolutionary" tenets of TOC, such as the evils of cost accounting and the overwhelming importance of constraints, have gradually been incorporated into the mainstream of management theory. The first book about TOC, *The Goal*,[7] came out in 1986. Dr. Goldratt had a difficult time getting it published, because it was a novel about serious business topics. It is now required reading in management programs at many colleges and universities. Slowly TOC seems to be coming of age.

Many TOC concepts have been applied with great benefit to project management. Many major corporations, including those mentioned above, are using these concepts today. In fact, there is every reason to believe that, despite its start in manufacturing, TOC can be even more powerful when applied to project management. Nevertheless, there are very few TOC books that are directly relevant to project managers. Some new management approaches that originated in manufacturing, such as Total Quality Management, have made the leap to project management; TOC still lags behind.

This book demonstrates some workable, common-sense solutions to vexing, chronic problems faced by project managers. We will try to demonstrate that the problems of late, over-budget, and under-performance projects are not inevitable. In Part I we work through the kinds of problems typically faced in project environments. Only after understanding the underlying nature of the problems, their causes and their effects, will we be in a position to develop practical solutions and have confidence that they can work.

There are many facets to really good, long-term solutions. We will need specific, improved techniques, such as reliable scheduling. Part II describes the Critical Chain scheduling method, which is a powerful means of gaining more predictability, productivity and speed from project plans. This technique is probably the most important new development in project scheduling in the last 30 years.

While TOC concepts can be applied to personal as well as organizational matters, the organizational context tends to be more difficult to implement, because there are many people involved. People must work together well to achieve substantial improvements. Much of this book deals with projects carried out in organizations. To be significant in this context, individual improvements are insufficient; they must be organization-wide. Local, one-person or one-area improvements are insufficient to enable an organization to keep up in today's world. Part III discusses necessary components of a global approach to management.

No ideas for improvement are more than amusing intellectual exercises unless they can be applied in real life. There are many obstacles associated with implementing TOC: resistance to change, insufficient understanding, counter productive measurements, and so on. A detailed implementation plan will generally need to be specific to a particular environment, but there are common threads. Part IV deals with these common threads and provides some pointers to success.

This book makes assumptions about what you do and how you work. As you read it, you will be making assumptions about what is meant. Many of these assumptions will not be obvious, but they can cause misunderstandings that block success in using the concepts. Consequently, you can't find here a substitute for thought. There are two fundamental rules to keep in mind as you read. First, **treat everything in this book as if it is wrong.** It's much more productive to spend time exposing and fixing old invalid assumptions than it is to create new invalid assumptions. Second, **treat everything you're currently doing as wrong.** Chances are you will need to re-think what you're doing. If you assume everything you're doing is right (and probably most of it is), this re-thinking will not be very productive. If you assume it's wrong, you may get some interesting ideas.

With these ideas in mind, this book will give you ideas and techniques that can help you to manage your personal and professional projects better, and understand what is happening and what to do when they go wrong.

Key Concepts

- Project managers are measured on schedule, budget, and performance.
- Projects are frequently late, over budget, and under performance.
- Common practice is not working.
- Theory of Constraints offers tools that can help.

Questions for Further Thought

1. What entries can you add to the cause–action table above based on your own environment?
2. If the things you're doing match ideas presented in this book, does that mean what you're doing is right?

Endnotes

1. "Additional Information Related to Improvements Needed in the Department of Transportation," Office of the Inspector General, May 1997, http://www.dot.gov/oig/statements/attach.html.
2. Document GAO/HR-97-9, "Information Management and Technology," U.S. Government Accounting Office, February 1997, 0:1.
3. Document GAO/HR-97-6, "Defense Weapon Systems Acquisitions," U.S. Government Accounting Office, February 1997, 0:1.
4. Report GAO/RCED-97-17, "Department of Energy: Opportunity to Improve Management of Major System Acquisitions," U.S. Government Accounting Office, Washington, D.C., November 26, 1996.
5. "Evaluation Results for 1994," Operations Evaluation Department, World Bank, Washington, D.C., 1995.
6. Touran, A., Bolster, P. J. and Thayer, S. W., "Risk Assessment in Fixed Guideway Transit System Construction: Final Report," Northeastern University, January 1994, prepared for the University Research and Training Program, Federal Transit Administration.
7. Goldratt, E. M., *The Goal, Second Revised Edition*, North River Press, Croton-on-Hudson, NY, 1992.

PROJECT MANAGEMENT TODAY

1 Bidding for the Project

Through the next few chapters we will be examining the process of planning and executing projects through a few stories. We need to understand the nature of the planning and execution processes from the inside. The reality is in what happens, not in what people have said should happen. The things that happen usually have much more to do with human behavior than textbook planning procedures, so we must take a practical, people-oriented point of view rather than a theoretical one. We must make the assumption that human behavior is rational and not arbitrary. We assume that there are knowable reasons for even the strangest behavior. If this assumption is true and we pretend it's not, solutions will not be targeted at real problems; they will appear to be whimsical and impractical. If it's not true, we're doomed anyway. We can't blindly accept the common explanations, because the common solutions aren't working. We must think things through. We will start at the beginning.

The planning process for any project starts with an idea — the idea of what the project should accomplish. If the idea is clearly stated, it can be called an objective. The idea could initially be defined outside the company. For example, defense contracts typically start with a "request-for-proposal" (RFP) that outlines what is needed and asks for bids. New building projects usually require bids from construction companies. Objectives also come from internal sources; as, for example, when the marketing department identifies a need for the engineering department to create a particular type of product.

Assuming the idea seems to have potential, the achievement of the idea must next be evaluated in terms of costs and benefits. This could be a short process, or it could involve significant expense by itself. Feasibility studies are frequently significant projects in their own rights. For some kinds of projects the evaluation could be ongoing, and a decision to cut the project could be made at any time.

We start with an example of the initial project planning that might be typical of a small defense contractor. The company is called "Televar," and the narrator is an experienced project manager named Janet.

Janet's Project

A couple of weeks ago my boss asked me to put together a bid for a 2-year project for the Air Force to design and build prototypes for a new wideband radar receiver. It's on the cutting edge of technology — exactly what Televar needs to be doing in order to survive. We're good at it, too. That's one of the reasons the Air Force asked us to bid. Nobody has more experience or a better track record than we do.

It's a shame that we only get a tenth of the bids we make, but somehow our costs are just too high. Our cost structure really cuts into our competitiveness; we're constantly losing out due to price. It's not that management hasn't been trying to downsize; counting losses of production people, we have only a third the number of employees we had 10 years ago. That's how it is in the military contracting business these days; you make drastic cuts just to keep the company afloat. But now we're stuck with even higher overhead ratios. It's a spiral that's sucked a lot of companies down, and we're starting to see the bottom of the tub ourselves.

Preparing the numbers for the bid was pretty standard. I went through the RFP with the technical guys. We broke the project into pieces and decided what kinds of resources would be needed, then they came up with more detailed task estimates. Of course, we didn't develop a detailed work breakdown structure, we just prepared enough detail to do the bid.

Naturally, we negotiated some padding for the tasks. Things are always changing and always going wrong, so if we didn't add some slack there would be no way of finishing the project in time. On the other hand the technical guys always ask for too much slack, so I have to talk them down. Unfortunately, the 24 month development period isn't negotiable, so we couldn't change that. After we finished I had to take out some time from the estimated task durations to get us there. This step gave us the basic data for overall project duration and resource requirements. Then last week I asked Roger in finance to put together some costs.

I have a feeling we can get this project if we can keep the bid under ten million. There's only one problem. This morning we met to go over the costs, and it didn't look good. With our overheads, the cost to do the project is $10.5 million without even adding any margin. We'd have to sell below cost to get the contract, which is out of the question.

I wasn't ready to give up without a fight; I know how important this project really is to the company. I asked Roger to recheck his figures for any places we can squeeze. Meanwhile I've gone through my estimates and trimmed some requirements. The new plan is much more optimistic; I only hope these cuts don't bite us later.

After I cut down the time for some of the tasking and Roger made some magic with the overhead allocations, we somehow got our costs down to $9.1 million. That doesn't leave us a big margin, but at least we can make my $10 million target and management will buy in. Now we just have to cross our fingers and wait for Uncle Sam to decide.

The Planning Conflict

Janet is struggling with the standard conflict when bidding for work: bid high in order to make sure the work can be completed on time and at a profit vs. bid low in order to get the bid.

In a perfect world, a task would require a certain known amount of time to complete. The job would cost some specifiable amount. The project would be profitable or unprofitable, we would bid or not. Very often in preparing bids we assume a perfect world. It's not a perfect world, for many reasons. For example,

- Task requirements can't be completely known until the work is actually done.
- Abilities and motivation can vary significantly between workers.
- Customers typically make changes which may (or may not) result in more profit.
- The final price leaves some margin to play with and avoid losing money.
- The costing process seems to inflate costs by arbitrary amounts, by allocating overhead expenses that may not have anything to do with the project itself.

In other words, nothing is certain. Janet made the following compromises to resolve the conflict:

- Lower "negotiated" times for each task
- Shorter task times across the board
- Decreased resource requirements
- Playing with the accounting numbers

From the compromises made we can deduce some things that Janet probably doesn't know:

- How long will a given task really take?
- How much slack is needed, where?
- How long should the project really require to complete?
- What will be the real bottom-line impact of the project on the company?

We pretend that the numbers drive our understanding. We pretend that the numbers are the cause, and the understanding the effect. The reality is the opposite. Because of the uncertainties, a basic part of the planning process must be intuition. Gut feeling is essential. The problem with gut feeling is that it is very difficult to communicate; its inherent imprecision makes it difficult to transfer to others. Meanwhile people without the intuition expect to see detailed, precise numbers. So the process of communicating the intuition requires changing the numbers. It's ironic that the numbers for the bid, which are a result of an attempt to express uncertainty, are not at all imprecise. They convey certainty. The certainty may be necessary to express to clients, but it does not express reality. Numbers expressing certainty are clearly a poor vehicle for communicating and evaluating whether to carry out a project.

Dealing With Uncertainty

Can uncertainty be eliminated? There are problems that can't be known until they are tripped over. There are interactions with other projects, there are differences between people, there is the impact of Wall Street on top management's actions, and on and on. Murphy's Law cannot be eliminated.

Uncertainty should be minimized and/or expressed. It should never be ignored. It must not be used as an excuse for not planning; it must become part of the planning. The purpose of a plan must be to quantify what we do know, so that we base decisions on as much real information as possible. In the process, we will also have to decide what we do not know, so that we can avoid erroneous assumptions.

Part II of this book describes one technique for representing and dealing with uncertainty, the "buffer." Buffers allow pooling of slack time to protect key areas of project schedules.

In the next chapter we will look at uncertainty some more, to get a different point of view on Janet's problem.

Key Concepts

- Project bidding and planning is frequently a matter of compromise and negotiation.
- Intuition is an inevitable part of the bidding process.
- Uncertainty is an important factor in project bids.
- Uncertainty should be expressed and communicated, not ignored.
- Project planning apparently involves compromises between uncertainty and market requirements.

Questions for Further Thought

1. Pick an industry that you're familiar with. In what ways is the bidding for new projects different from the example given, and in what ways is it the same?
2. What would be the value for Televar in keeping historical data on how long various kinds of tasks take?
3. What are some ways people attack uncertainty? How well do these ways work?

2 The Worker's Viewpoint

No project can be completed without someone to do the hands-on work. In different environments they might be engineers, computer programmers, roofers, or X-ray technicians. They might manage a team or be part of a team. The workers can have a critical role in the planning process. They have intimate knowledge of the work to be done. They estimate how long tasks will take, and they are responsible for meeting those commitments.

In this chapter we look at the view from lower in the organization. What influences people's behavior? What kinds of problems arise? Can they solve these problems alone?

Joe's Story

Every Friday Joe and his friend Paul, both workers at Televar, go to Henry's Pub to mull over the week's events. Joe is a senior engineer in the hardware design group. He manages teams and has a lot of administrative experience. Joe is outgoing but somewhat cynical; he has been around. Paul is a technician, less senior than Joe, part of the group that's responsible for making the designs work. He tends to be more thoughtful. The men enjoy getting together at the end of each week to talk shop and drink beer. Joe is sitting at the bar when Paul walks in.

"Sorry I'm late. I've got about four deadlines right now — it's really crunch time. Everybody in our group seems to be working overtime. I'll probably spend all weekend in the lab."

Joe laughs and says, "No problem. Things are pretty quiet with us. In fact, I kind of wish we had some more work. I don't suppose there's anything you want to farm out to us?"

Paul looks pained. "That sure would seem to make sense, wouldn't it? There's no way our budgets would allow for that. In fact, my boss claims we're over budget on most of our projects already. Don't worry, we'll wait until it's too late and then ask for help. That's standard operating procedure."

Joe smiles and replies, "Hey, no problem. Our motto is, if you've got a charge number, we're there for you."

Paul shakes his head ruefully, and both men pause to order beer. "So what's new with you?" Paul asks.

Joe jumps right in before the bartender is even finished pouring. "I'm spending most of my time on some internal R&D work. But one thing happened this week that you'll get a kick out of. On Monday I was given an RFP and asked to estimate how long it would take our guys to do some of the programming work. They need modifications to a pretty standard receiver tuning algorithm, so I put down six months."

Paul interrupts and says, "You mean like that Navy project you worked on last year?"

Joe says "Exactly. So ..."

"Wait a minute," Paul interrupts again. "Last year you told me someone worth his salt could do that kind of job in two months. Why all of a sudden six? Did you get worse at it?"

Joe takes the arrival of the beer as an opportunity to down half a mug, then says "Of course not. That was two months of real work. This is six months of elapsed time. Don't you know the difference?"

Paul says "I never have to deal with it. They give me the work, I do it. Sometimes I have a lot of stuff to do at once, but my boss is pretty good at shielding me from that scheduling nonsense. He knows I do my best. What's the problem?"

Joe thinks for a moment. "Well, there are lots of reasons we have to say six months. Let's see. First of all, you can never predict what will happen when you get into something. If I want to make sure I meet a time estimate, I have to increase the timing significantly." Paul looks puzzled, so Joe reaches for a napkin. "Here, it looks something like this," he says, and draws the graph shown in Figure 2-1.

He continues: "The horizontal axis is time; height of the curve represents the probability of finishing a job during any time period. The real distribution probably doesn't look like this, but it's the general idea that matters. You have a big peak toward the start, and a long tail on the right. That means the job can usually be finished fairly quickly, but sometimes unpredictable things happen that can slow it down a lot." Now Joe is getting warmed up. "Rookies estimate the mode of the curve, which is the top of the hump. That's the most obvious choice, but their chances of finishing in that time are quite a

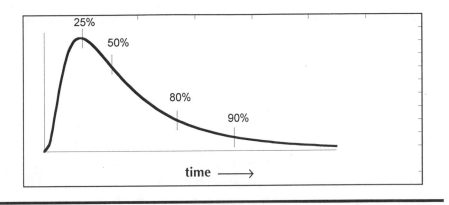

Figure 2-1   Lognormal Distribution

bit less than fifty percent. I marked the average as 50%, but if we estimated that, even if nothing else happened, we'd be wrong half the time. No, most of the time we pick a time way out here at eighty or ninety percent, so that most of the curve is on the left. Cover your ass. How well you follow schedules is important at salary review time."

Paul has been dividing his attention between the beer and the diagram. Having successfully negotiated the beer, he signals for another, then says "That looks like maybe twice as much time. You went from two months to six months. Where's the rest?"

Joe says, "That's right, the basic uncertainties are only part of the problem. There's the fact that different engineers will take different amounts of time to do the job. Also people keep getting jerked from one thing to another when crunch times come." Paul nods his understanding. "So six months is probably realistic."

Both are silent for a moment, musing about how a two-month job could take six months or more. Paul, after thinking hard, says, "So you have to overestimate, because otherwise you'll be late too often."

"Right."

"But," Paul continues slowly, "you couldn't possibly keep your job if you spent six months working only on one project, because then you'd look even lazier than you are. Right?"

"Right."

"So I bet your boss volunteers you for several overlapping jobs, which just about guarantees you'll never get two months to work on a two-month job."

"Uh — yeah, that sounds like it. Usually there are lots of things going on. That's another reason we have to overestimate."

"So," Paul concludes with a frown, "by this time nobody has the slightest clue how long anything is really going to take. How can you possibly finish anything on time?"

"It's worse than that. That's what I was going to say when you came in. The powers on high usually cut time from our estimates so they can make competitive bids to get new work. I found out today that my six-month estimate was pared down to five." He smiles. "Do you still think I overestimated?"

Paul, still frowning, says "So that probably explains why we have such bad crunch times, right? If nobody really knows when we're doing our work or how long it will take, how can you avoid big overloads? In fact, how can you avoid being late? Hey, I bet that explains why the design work we get from you guys has — uh — occasional mistakes."

Joe studies his beer for a minute, then says, "Probably sometimes we do cut corners in the rush to get things out. When you put it that way, it seems inevitable. But our on-time rate is pretty good. We really pull out all the stops to meet deadlines."

Paul starts laughing and says, "So our projects usually finish on time?"

Joe laughs too. "You've got me there. But the guys in my group are mainly measured on task completion, not project completion. We get a lot of pressure from project managers, but late projects aren't really our problem. Anyway, most of the time it's not us who make projects late." Then he becomes thoughtful again and says "What's interesting is, given all the stuff going on, if we're overworked sometimes we should also be underworked other times."

"Why is that?"

"If everyone is either fully loaded with work or overloaded, everything would get later and later. As time goes by, projects would get later and later. That doesn't really happen. And I've noticed that sometimes there really are periods of less work."

"Which is where you are now."

"Exactly. So we find things to do. We find ways to improve old designs, or slow down, or do things nobody cares much about. We always try to stay busy. Remember that layoff last year? It'll happen to you if you're caught napping."

"Yeah, we lost four guys. But somehow it doesn't seem like this makes any sense. Can it be right to fool ourselves about how long things will take? Can it be right to fool our clients? Can it be right to look busy to fool our bosses? It seems like that has to bite us."

"Well, here's how I look at it." Joe smiles and leans back against the bar. "If we made some kind of average time estimates, we'd be wrong at least half the time. We'd get fired. If we weren't willing to take on several tasks at once, we wouldn't keep busy. We'd get fired. If we did a bad job, we'd get fired. If we took the time to do everything perfectly, we'd be late, and then sooner or later we'd get fired. The only possible answer is compromise. You just do the best you can."

Paul stares sadly at the disappearing suds in his empty beer glass. Suddenly he looks up, smiles and points to the television over the bar. "Hey, look, the Red Sox! The World Series must have started!"

Discussion

Joe has described a basic part of the give-and-take of project planning. Typically, time estimates are inflated, whether by the project managers, resource managers, or workers. This is a response to the inevitable uncertainties that occur. In order to prepare competitive bids — that is, in order to take the external market into account — times are then frequently pared down so that the bid can be won.

This give-and-take process can cause much of the real information that could have existed in task-timing estimates to be lost. That, in turn, means that schedules must frequently be revised as they are time and again proven wrong.

It looks like a no-win situation. People must take on multiple tasks.[1] They must expand their time estimates. They must compromise quality. And they do this because of their measurements, the standard project measurements: cost, quality, and schedule. They must keep busy to cut costs. They must produce to some minimum level of quality. They must finish their tasks within the allotted times.

During project execution, the project managers also get into the act in many important ways. Project managers want their jobs worked on, because they themselves are measured on schedule conformance as well. They induce people to change between tasks. They don't want to go over budget, so they often can't bring in more people to work on a project, even if those people are doing nothing. Of course, they can't always tell when people have nothing to do, because people — especially engineers — can always find ways of staying busy.

These compromises often produce a more-or-less acceptable equilibrium. This is fine for companies that have other kinds of competitive edges, such as advanced technology, or for companies willing to settle for mediocre results. It is not acceptable if you want outstanding performance. It is not acceptable if you understand the magnitude of the lost productivity, or the kinds of competitive advantages being forgone. So far we have barely scratched the surface.

The next two chapters, 3 and 4, go into some detail describing the competitive advantages and productivity that are lost due to current practices. Chapters 5 and 6 tie the problems together, so that we can understand the connections between them.

Neither Joe nor Paul nor Janet is in a position to create significant change by him- or herself. Changes will require cooperation between groups throughout the company. In Chapter 7 we will start to look at the nature of the required solutions.

Key Concepts

- People do things for reasons, even if you don't understand or believe the reasons.
- Workers usually feel compelled to stay busy.
- Workers must meet task completion commitments.
- Task timings need to be long enough to allow for uncertainty.

Questions for Further Thought

1. Should we care if tasks are late? When?
2. To what extent are task duration predictions self-fulfilling prophecies?
3. What are the undesirable things that Joe and Paul experience?
4. Which of these undesirable things does it make sense to tackle in isolation?
5. Do you see any ways that Joe or Paul, by themselves, could be expected to improve the situation significantly?
6. How many of the problems we've discussed would occur in your environment?
7. Now what do you think would be the value for Televar of keeping historical data on how long various tasks take?
8. Suppose that a project is late, and it could be made less late if a particular resource were added. Suppose that resource is available to do the work. Should the resource be added to the project? What would be the real impact on the company as a whole?

3   Hidden Costs: Work-in-Process

In Chapter 2, Joe mentioned that he must make long task estimates, and that he must consequently accept work on more than one task at a time in order to stay busy. This practice is called "multitasking." In order to understand in detail the ramifications of multitasking, we must first dive into a concept familiar from the manufacturing world called "work-in-process."

Work-in-process (or "WIP") is work which has been started but not finished. Every project that is started but not finished or canceled has WIP associated with it. The pile of half-written or half-read documents sitting on your desk or nightstand, the partially built cabinets in the bathroom, the new computer hardware sitting on the floor — all are WIP. As we will see, the more WIP that exists in a system, the longer it takes work entering a system to leave. Excessive amounts of WIP can have a big negative impact on quality and time-to-market, which in turn can significantly affect ability to compete in most markets.

The Traditional WIP Tradeoff

In recent years much has been written about the negative effects of WIP in manufacturing, often under the banner of Just-In-Time or Zero Inventories. While these concepts are traditionally applied to manufacturing, many of them are at least as valid for projects.

In the project world, WIP may represent external investment, if there are vendors or subcontractors. It may also represent only internal kinds of costs: the time spent by people and machines to get the work to a certain point.

For example, a partially completed diagram on Joe's desk may be the result of a number of tasks, including a bidding and planning process, work with the customer to determine requirements, and research on available materials and processes. Project and resource managers seldom pay a lot of attention to work-in-process, for several reasons:

- It's hard to quantify WIP in a way that's meaningful.
- Some tasks are ongoing, requiring a commitment of a percentage of time rather than a fixed block; for example, membership on a committee. These activities are also hard to quantify as WIP, and instead are built into schedules (or, sometimes, ignored entirely).
- Some tasks, especially for internal projects, don't require a clear commitment to a delivery date. For example, commitments to review papers, develop tools, or provide education are often not concrete. They can apparently be ignored. They sometimes get in the way of more important tasks; they sometimes lie forgotten at the bottom of the pile.
- The negatives associated with high WIP are not well understood by most project managers.

Traditional theory addresses a tradeoff between the carrying cost of WIP inventory and its value in keeping people and machines busy. On the one hand, there can be significant investment tied up in the raw materials (and perhaps the productivity) that go into building WIP inventories. It seems useful to minimize this tied-up investment. On the other hand, people can't work without something to work on. Some amount of inventory is necessary in order to keep people productive. Furthermore, WIP provides a protective buffer in case things go wrong. If someone has work, they can continue being productive even if the work center providing them with tasks takes a vacation. In this way WIP represents a tug-of-war between costs and productivity. Consequently, traditional valuation puts WIP inventory somewhere between a liability and an asset.

The TOC WIP Tradeoff

Theory of Constraints (among other more recent management approaches) looks at a somewhat different tradeoff. On the minus side, increased WIP levels imply an increased time between material entering a company and actually reaching a customer. For example, a typical U.S. car dealer keeps a month of inventory on the lot. That means it will always be at least a month from the time a new model is produced at the assembly plant until it actually reaches consumers; the pipe is clogged with old models. Since consumers know when the new cars are coming, at the end of a model year dealers have little choice but to unload their current inventories at bargain prices.

WIP and the corresponding lead time is everywhere. Have you ever walked into a restaurant, only to leave immediately (or worse, after waiting around) because the wait was too long? The "lead time," or time before you get your table, is too large. A shorter lead time than the competition's can result in more sales, often at higher prices, in many types of business.

Reduced inventory gives a company a competitive advantage by allowing it to satisfy customers more quickly, both by reducing the time between order and delivery and by reducing the time in which new products can be introduced to the market. That reduced lead time also allows quality problems which are due to an ongoing problem (such as defective testing equipment) to be discovered more quickly, because customers are using the products more quickly. The quality problems therefore affect fewer products. Furthermore, the causes can be discovered more easily, since they're closer in time to the bad effects.

In project environments, short lead times mean a reduction in changes or cancellations of orders, because customers have less time in which to change their minds. The longer a project lasts, the more likely its requirements are to change. This happens because of advances in technology, especially in high-tech fields, and also just because customers and designers have more time to get new ideas. The more quickly a project can be completed, the less time it spends as WIP, the fewer changes will be required.

A lower WIP inventory level is beneficial in many other ways. It can lower confusion by allowing simpler schedules. It can decrease the stacks of papers to be sorted through, making locating key documents and information much easier. It can increase the predictability of deliveries, reduce reliance on questionable forecasts, and on and on.[2] By itself, a reduction in work-in-process can be a huge competitive advantage, and hence a powerful lever for improvement.

On the plus side, some work is still essential in keeping resources busy. There must be something present for people to work on. You can't complete anything on time if you can't start it; and it certainly doesn't look good to be idle. But as we will discover in Part III of this book, not everyone needs to keep busy, because the overall productivity of a company depends on only a very few resources.

In some ways, work-in-process is similar to the water in a sink. In Figure 3-1, work is coming into the system through the tap at the top. The fluid filling the sink is work-in-process. The vertical height of the fluid is proportional to how long the sink would take to drain. There is a long wait between the time any new work enters the system, and the time it reaches customers through the bottom. The star represents a new product. It will have to wait until it gets to the bottom before it can be introduced to the market or shipped to a customer. The circle with a line through it (a "No" symbol) represents the effects of a quality problem. It is bad work, which may not be discovered until late in the process, either during system tests or when the project reaches the customer. By the time the problem is discovered, it's possible that the engineer who created it and could diagnose it has left the company. Now consider the corresponding low-inventory case. In Figure 3-2, work-in-process is comparatively low, so the lead time is much less. New products can be introduced quickly, and the time until the quality problem is discovered and fixed is much shorter.

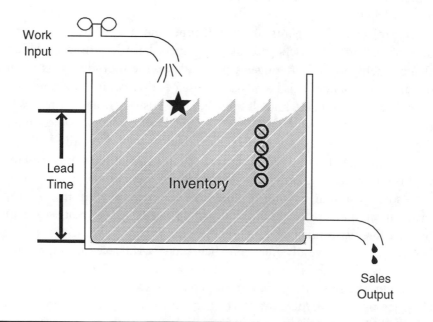

Figure 3-1   High Work-in-Process (Bad)

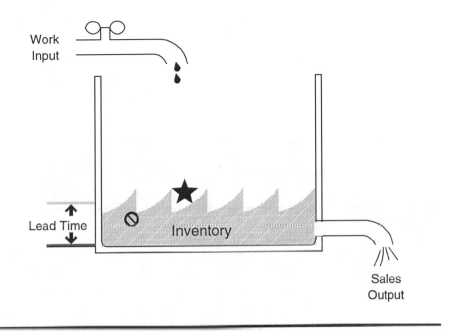

Figure 3-2   Low Work-in-Process (Good)

Working on Large Batches

A common way that inventory and lead times are artificially inflated is by transferring large batches of work between resources. In Figure 3-3(a), the bottom resource does a large batch of work, then transfers it to the middle resource, which completes it and passes it on to the top. This work could be, for example, a set of four drawings that are first designed, then drafted, then passed on to the customer for review. In Figure 3-3(b), the drawings are passed along one-by-one. Notice the difference in time it takes for the customer to sign off on the entire set. Suppose the customer discovers a problem with the middle resource's work. In which case will fixing the problem cause the least waste? Figure 3-3(a) is a high-inventory, long-lead-time environment. Converting to Figure 3-3(b) can result in a substantial competitive advantage.

Notice that the problem shown in Figure 3-3 has nothing to do with the actual durations of the batches. It has to do with how work is transferred. It isn't necessarily bad to do large batches of work; in fact, in Chapter 4 we'll see several reasons why small batches can be bad. The problems occur when

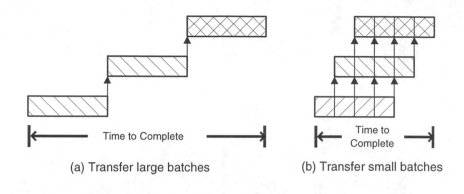

(a) Transfer large batches   (b) Transfer small batches

Figure 3-3   Batch Sizes

the large batches are viewed as indivisible. The distinction must therefore be made between "transfer batches," which are the chunks in which work can be transferred between workers or work centers, and "process batches," which are the discrete chunks of work done by workers or work centers.[3] For example, the architectural drawings for a new facility might be considered a process batch. The set of drawings is conceptually a distinct part of the construction project, and it might be done by one architect. Pieces of the set of drawings (transfer batches) can be given to the builders, so that construction can be started before the entire set of drawings is complete. Smaller transfer batches are usually better, because they decrease lead times. Longer process batches are not necessarily bad.

Very often tasks are broken out and transferred according to divisions suggested by the standard project life cycle approach.[4] That approach has the following kinds of elements:

- Define the problem being addressed
- Analyze solution requirements and feasibility
- Design the solution
- Develop the solution
- Implement the solution
- Utilize/maintain the solution

This is a useful model for many reasons. But it can also lead one to plan projects in large chunks that are not overlapped, rather than small tasks that can be overlapped. This causes larger transfer batch sizes and longer project durations, with drawbacks that we've just discussed. For example, consider a project which has several pieces that need to be designed and then developed. The project duration will be much less if the design and development phases are split into several small tasks and done as much as possible in parallel, rather than just having overall design and development phases for the project done in strict sequence.

Keeping Busy by Multitasking

Frequently formal and/or informal measurements conflict directly with a low-inventory policy. We've seen that already with the need to estimate long task times. Longer task times mean that projects stay "in process" a longer time, meaning more WIP.

In Chapter 2, Joe talked about committing to multiple tasks in order to stay busy. There are many people who juggle far more tasks than they can actually keep in the air without superhuman effort. It's a rare person who does this and doesn't drop balls. When they do, we tend to be forgiving: dropping one or two out of ten doesn't seem so bad. Keeping busy, even to the extent of overcommitting, is perceived as good.

Overcommitment can provide other perceived advantages to the worker besides the ability to stay busy. Being overworked is an excellent excuse for almost anything. It becomes very difficult to criticize someone for not making progress on task X when they are also responsible for tasks A through W. And finally, who is more likely to be perceived as a valuable worker: someone who is constantly juggling ten important tasks, or someone who is either working on one task or waiting for their next task?

The most common form of overcommitment is **multitasking**. Multitasking is accepting (or being given) more than one task to be worked on concurrently. Multitasking may seem like a good way of keeping people busy. In reality it often combines the worst features of large batch sizes and overcommitment, since generally all tasks will take much longer than if they were done one by one. If you have two tasks to work on concurrently, each requiring a week to complete, it will likely take you at least 2 weeks to finish either. Often the tasks with the most inherent risk will be finished last, which is also dangerous. Furthermore, more than 2 weeks may be required if switching between tasks takes time. This is called "setting up," and we'll discuss that in Chapter 4.

Nevertheless, it is frequently not possible to assign people to only one task. There might, for example, be resources that are indispensable to more than one project. The worst problems with multitasking come about when people switch frequently between several tasks, because then all tasks will take much longer than they need to. This means that when multitasking is done, the priorities between tasks should be very clear. That way the highest-priority task will finish as quickly as possible. If for some reason a person must wait to continue a higher-priority task, then — and only then — should they put in time on the lower-priority one.

The conflict inherent in high WIP will be solved in detail in Chapter 10. For now, we'll just reassert what we said above: it is not necessary to protect productivity everywhere. The demonstration must wait.

Key Concepts

- Work-in-process (WIP) is work inside a system that is started but not complete.
- There is a traditional conflict between wanting to lower WIP levels to reduce the carrying cost and wanting to raise WIP levels to maintain higher levels of productivity.
- A more valid conflict is between lower WIP levels to improve ability to compete and higher WIP levels as a kind of insurance, both to maintain productivity and to be able to finish tasks on time.
- High levels of work-in-process can cause many problems, including longer lead times and confusion.
- Large, nonoverlapped batches of work frequently increase lead times in project environments.
- Always be on the lookout for ways to shorten lead times by breaking large batches into smaller batches, and then doing work in parallel.
- Multitasking, a common way of staying productive, can increase lead times significantly.

Questions for Further Thought

1. Calculate how many weeks or months of work you would have if you accepted no new jobs, but only worked on things already in process at home or at work. Is this time a reasonable measurement of your work-in-process?

2. Think of three areas in which you frequently experience a buildup of WIP. Would your actions change if you used work-in-process as a personal measurement? How?

3. How do Joe's actions in Chapter 2 cause a buildup of work-in-process? What is the likely impact on lead times?

4. Does your work environment encourage multitasking? If you must work on several tasks at the same time, are there clear priorities between them? If not, why not?

4 Hidden Costs: Lost Productivity

We have all seen or heard of examples of the amazing productivity that can be achieved when people are focused on urgent, important work.[5] Contrariwise, a lack of priorities and arbitrary measurements can sap intensity and drain productivity, often in nonobvious ways. Common practice, including the need to keep busy and the need to adhere to firm task commitment dates, usually tends to support the latter situation rather than the former.

Since we'd prefer to prevent lost productivity, we need to understand the specific mechanisms through which it is lost. This chapter describes many of them. These mechanisms reflect lost opportunities rather than directly quantifiable numbers. The magnitude of the losses is important to understand so that we can see the extent to which productivity is hurt by common policies. We might imagine that it could be significant, but *how* significant is impossible to understand before it happens. The precise magnitude of these losses isn't really important; they can't be quantified precisely anyway.

Parkinson's Law

As we've seen, people can stay busy by multitasking. They can also cause work to expand to fill the available capacity (Parkinson's Law[6]). If I don't know of any new work coming in next week, I may slow down what I'm doing this week. If I have one task due in 3 months and I know it shouldn't take me 3 months, I may start very slowly. I may hesitate to declare it done. Parkinson's Law is a major cause of lost productivity, and it's almost invisible. In fact, it's much more pervasive than most people believe. Frequently the people obeying it don't realize what they're doing.

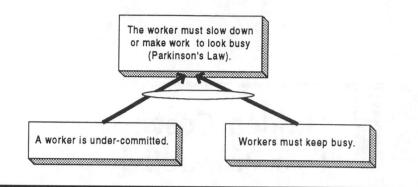

Figure 4-1   Cause–Effect

How real is Parkinson's Law? Let's approach it rigorously, in terms of cause and effect. We say that people want to stay busy, and sometimes they don't have enough to do. In that case, we claim that they will find things to do — Parkinson's Law. This logic can be diagrammed as shown in Figure 4-1. The boxes at the bottom are our two causes, both of which have arrows to Parkinson's Law. The arrows point causes to effects. The ellipse binding the two arrows means "and." We read **if–then** to indicate the causal link. The diagram therefore reads "**If** a worker is undercommitted, **and** workers must keep busy, **then** the worker must slow down or make work to look busy (Parkinson's Law)."

This seems reasonable, but let's check it further. What kinds of effects are we expecting if Parkinson's Law exists? In other words, how can we fill in the top box of this picture. Consider Figure 4-2.

Let's postulate some things we might expect to see if Parkinson's Law is valid:[7]

- People are almost never idle.
- During sudden periods of critical need people suddenly become more productive.
- Time is often wasted at the start of a project (also called the Student Syndrome[8]).

None of these effects "proves" the cause; each one adds more evidence. Each predicted effect that really exists helps validate our cause; any one that doesn't can invalidate it. Do you think it likely that Parkinson's Law holds in your work environment? What about at home?

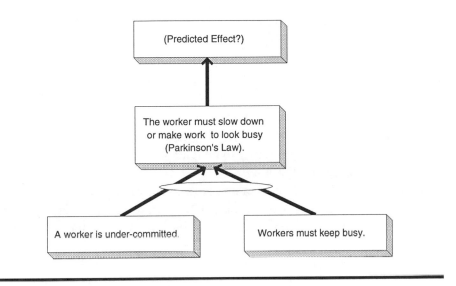

Figure 4-2   Effect–Cause–Effect

Parkinson's Law can become an invisible part of the corporate culture, obeyed by managers and workers alike. How can it be attacked? A clue is shown in Figure 4-2. If we can either keep workers from being undercommitted or take away the need to keep busy, the cause will be removed. Remember, both causes are required to produce the effect (Parkinson's Law). Removing one means that the effect will disappear as well. As Joe said in Chapter 2, undercommitment is a necessary consequence of overcommitment. In Part III of this book we will see further reasons why it's not possible for very many resources in an organization to keep busy. As we'll see in Chapter 5, there are other negative effects stemming from the need to keep busy, so it makes sense to attack that cause.

Setting Up

When there is multitasking, people are frequently switching between tasks, because they must show regular progress on each task. Most people require a certain amount of time to switch between tasks. The time needed to switch between tasks is called "setting up." There are both obvious and hidden costs to setting up.

The obvious costs are the direct time people require to find the relevant materials, refamiliarize themselves with the ideas, and in general get mentally and physically prepared to work on the task.

There are hidden costs that have to do with keeping abreast of multiple projects and technologies. If someone has several tasks to do concurrently, they must keep abreast of developments relative to these tasks. They need to know more than just when their task is ready to start; they must keep up with changes to specifications, guidelines, programming environment, and so on. The time span over which they must keep up with the changes depends on the elapsed duration of the task, rather than the amount of work the task requires. This can add significantly to people's workload and can be a significant hidden cost to multitasking.

Work on the tasks at hand can also slow down when one has other tasks in mind. One's thoughts can often wander between responsibilities, especially if the responsibilities are important or interesting, thus wasting time for all of them.

Using More People

In Chapter 3 we discussed the negatives associated with large batches of work. It makes sense to split work into small transfer batches, so that small amounts of work are transferred to the next resource. This decreases lead times. However, it may not make sense to split up the work of a task among several people; that is, to change the task into small process batches. Setups will increase. Furthermore, since the tasks are closely connected, the need for people to communicate can be increased significantly; "Men and months are interchangeable commodities only when a task can be partitioned among many workers *with no communication among them.*"[9] And, of course, there is no point to splitting a task if it must be performed sequentially; "The bearing of a child takes nine months, no matter how many women are assigned."[10] This means it's not always possible to add more resources to a task and expect it to take less time. There quickly comes a point where adding more people actually increases the time it takes to do a job, through increased communication requirements.

Mañana

There will always be work to do tomorrow, there will always be tomorrow in which to do work. The "mañana" attitude implies that work will finish when it finishes. There's no hurry on today's work, because tomorrow's work will always come. It is caused by the lack of a real incentive to complete something quickly. The competitive runner wants to win; he saves a kick for at the end of the race to sprint through the finish line. The exhausted or demoralized runner does not have this energy. He loses a sense of urgency.

Working on a seemingly endless pile of tasks, a pile that is added to as soon as something is finished, can seem like running up a very fast down-escalator. If you even pause to catch your breath you're farther behind. There is little satisfaction in completing any individual task, because there is no sense of having arrived anywhere. The pile doesn't go down, it may never go down. This kind of high-WIP situation can be very demoralizing. It can discourage people from having the drive, the enthusiasm, the kick that moves work ahead quickly.

Starting or Finishing Early

Suppose Joe has a task, and he knows its scheduled start and end times. So far the project is on time. Let's look at two cases. In case 1, he has several other jobs to do at the same time; in case 2, this is his only task. Under what circumstances will he finish early?

First, Joe knows that the people who need his work are unlikely to want it early. They have no reason to start it before their scheduled start time, which means effort spent in finishing early will be wasted.

If he has other jobs to do, and sees himself finishing early, there's a good chance he will devote more of his time to the other tasks. After all, he has to finish them on time also. The penalty for being late is much greater than the reward for being early. This is multitasking at its worst; everything takes a long time to complete, and nothing completes early.

If he has no other jobs to do, he doesn't want to finish and be left with nothing to do. So he slows down, maybe goes to meetings, takes a vacation, cleans his desk. The work expands, filling the time available.

The answer is, Joe will almost never finish early, because it won't help anyone. If Joe knew that the people his work goes to would like to start early, then he might consider finishing early. If the project is behind schedule he will probably finish as quickly as he can — unless, of course, he has other urgent tasks as well. The schedule dates become a self-fulfilling prophecy, and a best-case prophecy at that.

It's possible that Joe might start a task early, if the person before him finished early, and if Joe has nothing pressing to do. The chance of both of these happening at once is small. The probability of this happening in a project, to the extent that the last task finishes early, is just about zero. We have to wonder how projects can ever finish early. And if they can't finish early, is it any wonder that they're always late?

Key Concepts

- Work expands to fill the time available because workers want or need to keep busy.
- Setups associated with multitasking can cause significant lost productivity, due to the obvious costs of switching tasks and the hidden costs of being mentally prepared for all the tasks.
- When a pile of work builds up, the mañana attitude, "tomorrow is another day," becomes hard to avoid.
- Lost chances to start or finish tasks early help ensure that projects cannot be finished early.

Questions for Further Thought

1. Does Parkinson's Law manifest itself as procrastination? Answer the question now!
2. What kinds of tasks do you like to put off until tomorrow? What kind of incentive does it take actually to accomplish them?
3. Try to think of some ways that frequent setting up can lead to errors.
4. Think of a few high-WIP situations that cause you to want to look elsewhere for something to do.
5. How does the inability to start tasks early relate to Parkinson's Law?
6. Under what circumstances should workers be given task start and finish times?

5 The Generic Current Reality

Now that we've discussed in detail some of the hidden costs of the ways Joe and Janet do business, it is possible to express some of the problems from Chapters 1 and 2 in a more rigorous form, called a "Current Reality Tree" (CRT) by TOC practitioners. This is a tree-like diagram that lays out and connects causes and effects.[11] There are a couple of major reasons for creating a Current Reality Tree.

First, the Current Reality Tree expresses the causal links that exist in our reality in a form that's possible to scrutinize and clarify. This form helps us better to understand the causalities we're experiencing. It also gives us a better chance of communicating the problems. All this is helpful as we attempt to find a solution.

Second, using a Current Reality Tree we can identify what are typically very few "core drivers." These are entities toward the bottom of the logical tree that cause most of the problems higher up. If you can eliminate a right core drivers, you eliminate all the problems that stem from them. If instead you try to eliminate the problems individually without eliminating a core driver, the problems will persist.

Each page of the Current Reality Tree in Figures 5-2 to 5-4 should be read from the bottom to the top, starting with entity 1, using if–then logic. For each box, evaluate whether the statement makes sense. Is it valid? Is it significant? The arrows represent causalities; the entity at the base of an arrow "causes" the entity at the head. Read the boxes connected by arrows, using "**if** ⟨starting entity = cause⟩ **then** ⟨ending entity = effect⟩."

For example, Figure 5-1 should be read "**if** workers are measured on how busy they are and whether they meet their commitment dates, **then** workers

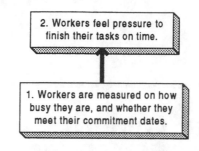

Figure 5-1   Reading the Arrows

feel pressure to finish their tasks on time." Evaluate the if–then statement carefully. Does the effect exist? Is the connection clear? Is the effect a result of the cause? Are there other important causes that are missing? The entire Current Reality Tree is read this way, making sure that all the arrows and all the boxes are read.

The numbers in the boxes have two purposes. First, they are useful labels to use when referring to the entities. Second, we have tried to make the boxes in this book convenient to read in ascending numerical order. Since all the arrows must be examined, this numerical sequence can't always be kept. In Figure 5-2, the causalities "**if** ⟨box 1⟩ **then** ⟨box 2⟩" and "**if** ⟨box 1⟩ **then** ⟨box 6⟩" should both be examined. Boxes in bold type are taken from previous pages, as with box 8 in Figure 5-3.

Ellipses that tie together arrows are read as a logical "and," as in Figure 4-1: all the tied-together causes are needed in order to achieve the effect. For example, the connections between boxes 5, 6, and 7 in Figure 5-2 should read "**if** workers cannot expect to keep busy with one task **and** workers try to keep busy, **then** workers accept, or are given, more than one task."

The Current Reality Tree, especially a "generic" one, is not a proof. It is a device that can improve understanding of causalities and problems, and help to communicate them. If an entity or connection doesn't make sense, see if you can add or change entities to improve it. You're encouraged to read Figures 5-2 to 5-4 now; we also include a text discussion.

We start with the assumption that workers are measured on how busy they are, and whether they meet their commitment dates (1). Workers might in this case be engineers, computer programmers, or carpenters — anyone whose primary responsibility is carrying out tasks in a project. The commitment date measurement implies that they feel pressure to finish tasks on time (2). But since things go wrong (3), that means they (or their managers) have

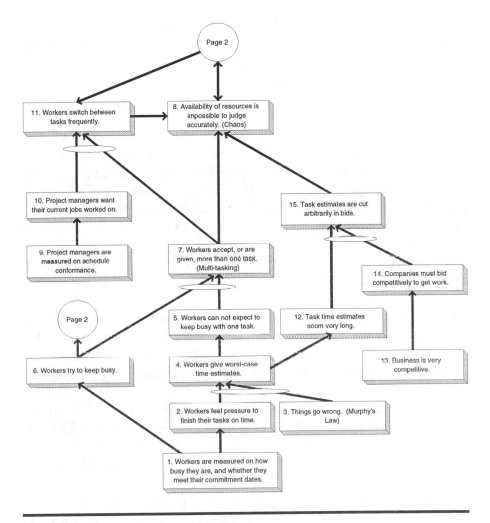

Figure 5-2　Generic Project Reality (page 1)

to give worst-case estimates for their tasks. In fact, the same logic holds for resource managers.

When workers give worst-case time estimates, they can't expect to keep busy with just one task (5). But since they try to keep busy (6) due to their measurements, either the workers or the resource managers will make sure they have more than one task to work on (7). That in turn makes it difficult to judge their availability. When we add all the causes, that difficulty becomes impossibility (8).

Going up the left side of Figure 5-2, project managers are also measured on schedule conformance (9). That means they don't want things to slip;

they want their current jobs worked on (10). But since workers typically have more than one task, in order to keep the project managers happy (an assumption not in the tree) they must switch between tasks (11). Workers switching between tasks adds to the difficulty of figuring out how much time the resources have available (8), both due to the general chaos (what are you working on?) and the need to set up frequently.

Now we go up the right side of the tree. If workers (or resource managers) give worst-case time estimates, task time estimates will seem very long (12). On the other hand, business is very competitive (13), meaning that companies must be competitive in their bids (14). But the combination of needing to be competitive and having long time estimates means that task estimates must often be cut in the bids. This, in turn, adds to the impossibility of judging the availability of resources.

There are several factors that can push management to underestimate resource availability, such as multitasking and switching between tasks. Parkinson's Law also contributes, (as we'll see in Figure 5-3) as well as other factors not included in the tree. On the other side, various competitive pressures can push management to overestimate resource availability. The net result is that it's impossible to know with any degree of accuracy how much capacity there really is to do work. A large degree of this uncertainty is man-made.

It's important to note that so far, nothing is anyone's "fault" except the measurements themselves, which are after all common — some might even say "best" — practice. Given these measurements, everyone is behaving exactly as we expect them to behave. In fact, we could add to the tree the assumption "People want to do their best."

From Figure 5-3, if availability of resources is impossible to judge, then (as Joe indicated) we must have both under- and overcommitted workers (16 and 17). Otherwise, if they're always overcommitted, things will continue to get later and later. If they're always undercommitted, the "problem" is solved quickly with layoffs.

If workers are sometimes overcommitted, then inevitably tasks will sometimes be late (18). This in turn implies fire fighting, which loops back to switching between tasks (11). This feedback loop between (8), (16), (18), and (11) can cause serious damage. There are also other effects of (16) and (18), which we'll see in Figure 5-4.

When workers are undercommitted, assuming as in (6) that they must keep busy, we expect them to slow down, as discussed in Chapter 4 (19). This is Parkinson's Law. But if the undercommitted workers are slowing down to

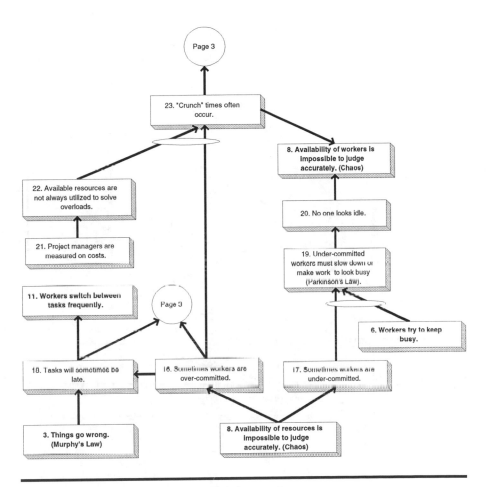

Figure 5-3   Generic Project Reality (page 2)

look busy, no one is going to look idle (20) which again loops back to (8), the impossibility of judging the availability of resources. This is an important feedback loop, which cements the impossibility of knowing for sure what the available capacity is.

On the other side, project managers are measured on costs (21). Budget is key; it has a direct and immediate impact on the organization's bottom line. But the way budgets are typically calculated often means, as Joe and Paul discussed, that available resources are not always utilized to solve overloads (22). Utilizing available resources usually counts against a project manager's budget. If sometimes workers are overcommitted, and even available resources aren't utilized to solve the overloads, there will be frequent "crunch" times as project personnel struggle against the overloads by themselves. These

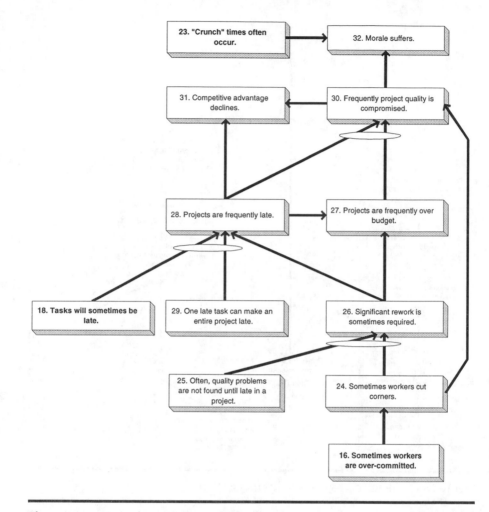

Figure 5-4   Generic Project Reality (page 3)

crunch times can add significantly to the general chaos, as seen by the loop from (23) back to (8) in Figure 5-3.

In Figure 5-4, when workers are overcommitted, they sometimes (some might say often) cut corners. Cutting corners might mean direct omissions, but frequently it's as simple as not checking one's work to the degree of detail necessary to find most of the problems. Most often cutting corners requires sacrificing quality standards. Very often these kinds of quality problems aren't discovered until late in a project (25), such as when components are integrated. In these cases significant rework is often required (26). This by itself can cause significant budget problems (27). In fact, "significant" probably doesn't convey nearly the extent of the disaster. In a study of 63 projects at TRW, Boehm found that an "error [is] typically 100 times more expensive to correct in the maintenance phase on large projects than in the requirements phase."[12] It's worth making sure things are done right up front.

Rework can also make projects late (28). In addition, given that one late task can make an entire project late (29), and we know from box 18 that tasks will sometimes be late, we also expect projects to be late. Notice that this lateness is made even more likely if tasks are typically not started early, because then only task lateness, never earliness, will propagate.

Rework requires additional resources, which can make projects go over budget (26 to 28). In addition, late projects can cause budgets to be exceeded just by taking resources for a longer-than-expected time (28 to 27). And naturally, if projects are late and over budget, the only compromise left is quality, whether that means reduced scope or unsolved problems remaining in the product (30). Short-term a reduction in quality can help reduce the lateness or cost overruns. Quality problems also stem directly from (24), workers cutting corners. Late and bad-quality projects also affect competitive advantage (31) (assuming the clients have alternatives, which is not always the case).

We could tie together many of the boxes in Figure 5-1 to point to bad bottom-line results; that is left to the reader.

Most workers care about what they're doing. It doesn't make people happy to put out a poor product. If an inferior product is shipped, morale suffers (32). Of course, there are many other reasons embedded in the tree that can cause morale to suffer. For example, frequent or long-lasting "crunch" times (23) can lead to low morale and burnout.

Core Problems

The "core drivers" or "root causes" in a Current Reality Tree are those entities in the tree that have no arrows leading into them. In the "Generic Project Reality" tree, these would be boxes 1, 3, 9, 13, 21, 25 and 29. Not all core drivers make sense to attack. For example, it probably won't make sense to attack box 9, "Project managers are measured on schedule conformance." Project managers need to bring in projects on time. The "core problems" are defined as those few core drivers that we choose to attack, and that cause the majority of the problems. They are the leverage points that can produce the most improvement with the least effort. Take a moment and analyze these entities to see which core drivers make sense to attack.

As an exercise, choose one of the boxes in Figures 5-2, 5-3, or 5-4 that contains a negative statement (e.g., 27, Projects are frequently over budget) and see what kinds of actions you would take to address it directly. This would probably be the same kind of actions as those we saw in the table showing cause and action in the Introduction. For example, to address the problem "morale suffers" you might have higher salaries, employee recognition programs, or other kinds of perks. Now think about the effects of your actions on the system as a whole.

The actions suggested to address the morale problem might have some positive effects; at least they might help keep people around. However, as long as the causes and causalities are still in place, the problem hasn't been "fixed." One effect may have been mitigated, but it might come back, and other undesirable effects still exist. We can't fix the problems just by attacking the leaves of the tree.

Some boxes are clearly not feasible to attack. For example, to attack box 13, "business is very competitive," we'd probably have to move to another planet!

Some boxes just seem wrong to attack. If we attack (let's say) boxes 9 and 24, measuring project managers by cost and schedule, there are still problems in the tree that won't go away. Furthermore, it seems sensible to measure project managers on costs and schedules. If we eliminated those measurements, we might cause worse problems.

Box 1, workers are measured on how busy they are and whether they meet their commitment dates, is clearly a promising candidate for a core problem. Its effects spread throughout the tree. By eliminating it, we reduce or eliminate many of the problems.

The other we'll choose to attack is 3, Murphy's Law. We can't eliminate it; uncertainty will always exist. However, if we can find better ways to deal with uncertainty, and if we can eliminate box 1 as well, we should be on the road to significant improvements.

If it makes sense, why haven't these entities been attacked successfully long ago? So far we don't have adequate techniques for dealing with Murphy's Law, and we have nothing with which to replace the local measurements for workers. In other words, either the cure isn't obvious, or it looks worse than the disease. As we proceed we'll see that TOC provides techniques for dealing with uncertainty and with measurements, techniques that allow us to generate major improvements. The cost of these improvements is willingness to change.

Key Concepts

- There are understandable causes for the problems we experience.
- Most of the negative effects in an environment are not due to people being bad, but instead are due to the policies and measurements that exist.
- Typically a large number of undesirable effects will be caused by a relatively small number of core drivers.
- Typically, eliminating a very few core problems can result in a huge improvement.
- We shall attempt to remove the undesirable effects (late, over-budget and under-performance projects) by attacking box 1, individual measurements, and box 3, uncertainty.

Questions for Further Thought

1. Think about an environment in which multitasking is not allowed. How could people stay busy? What are the likely effects of them staying busy in this way?
2. Suppose a key measurement for workers is the volume of the screams from the salespeople. Would this promote multitasking? How would it fit into the tree?
3. Revise the Generic Project Reality tree to reflect the specific situation in an organization you're associated with.
4. Try to identify all the loops in Figures 5-2 to 5-4. How does the feedback make things worse?
5. Frequently, when projects are late, more resources are applied. Of course, this makes shortages more acute, increasing the likelihood that other projects will have problems. How would you draw this additional loop?
6. How could you insert the statement "Productivity decreases" into the tree?
7. Reread the stories from Chapters 1 and 2. Which areas were covered in the current reality tree? Which areas weren't?

6 The Project Manager's Viewpoint

Before moving on to solutions, let's look some more at the human factors in project execution. The resources that are needed to complete projects aren't just chess pieces that can be moved from one part of the board to another. Frequently we experience problems that seem far in excess of the material considerations. In some cases workers and management are at each other's throats.

To see the kinds of things that can happen, let's look at a situation that could occur much later in a project. Janet is managing the project she earlier bid on, but things aren't going well.

Monday

It's only noon on Monday and already I feel completely used up. Things started to go downhill early. I had just gotten a cup of coffee and sat down at my desk, trying to make sense of our current schedules. Manny, my boss, walked into my office, sat down and said "Janet, we've got a problem with the XR-15 project. Jim tells me that he has a complete roadblock, and your guys are the only ones who can get it unglued quickly. Sorry, but he's way behind schedule right now, and any extra delay is likely to cost us big. I need to loan him Mike and Steve for a couple of days."

Mike and Steve are two of my best people. What does "a couple of days" mean, anyway? I'd bet money I won't see them around here for at least a week. It wouldn't be so bad, but we're starting to run into some problems in the wideband receiver project. I know that if this project is late, it'll be my fault no matter how many staff members are "loaned" away. I told Manny my concerns; at least there's no problem talking with him. It also doesn't do much good. He said there's nothing he can do.

Of course, this is no surprise; it happens all the time. And to be fair, my boss isn't the only one who's a wimp. It seems like every time a project gets in trouble, people start running around moving people and equipment from one place to another. The mentality is to put everything off until tomorrow, then panic. High-level managers seem to spend half their time creating fires and half their time putting them out. Of course, that means no one trusts their boss. Project managers don't trust each other much, either, because you never know who's going to take away your key people next. The rank-and-file people working on the projects don't know what they'll be working on next; just when they get going on something, it changes. Sometimes they have several jobs at once, all highest priority. You can hardly expect them to trust their managers. And when things get tense, which is most of the time, the arguments are unbelievable.

Somehow we lose sight of the fact that we're all supposed to be working together, for the same company. I don't know the answer, but I know there has to be a better way. So I sit at my desk, holding my head in my hands, debating whether I should have lunch, wondering what to do.

Current Reality

Janet is experiencing problems that can destroy morale and burn people out. They are not uncommon problems; Figure 6-1 shows an example of how they can come to pass. Boxes from the "Generic Project Reality" tree, Figures 5-2 to 5-4, are printed in bold type. Once again, start with boxes having no arrows entering them and check their validity. Read the connections as "if–then," and again check the validity of the statements.

Discussion of "Use of Resources" Tree

Project managers want their current jobs worked on (10); but we know from Figure 5-2 that sometimes workers are overcommitted (16). That means project managers don't always get the resources they want (100). This can be made even worse by the loop suggested in Question 4 from Chapter 5: when projects are late, they take resources from other projects.

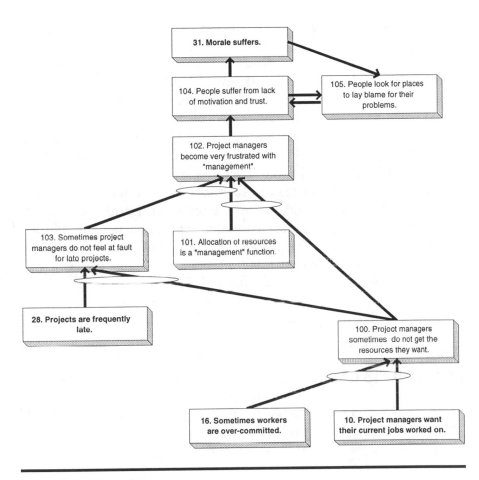

Figure 6-1 Use of Resources

Generally, allocation of resources is a "management" function. Often higher-level management will have to decide which project gets priority. Since project managers sometimes don't get the resources they want, there are sometimes bad feelings toward their management.

Furthermore, if projects are frequently late, and project managers sometimes don't get the resources they want (and sometimes need), then sometimes project managers don't feel at fault for late projects (103). It isn't their fault that their resources were pulled away. This is a problem Janet alluded to in the story above. Of course, that can also cause project managers to blame "management."

We can say, in general, that if project managers are frustrated with management, they and their management suffer from feelings of frustration and helplessness, which often manifest themselves in a lack of trust (104). There's also a lack of motivation, because it's hard to be motivated to work for someone you don't trust. Human nature implies that people will look for places to lay the blame for these problems (105). Naturally there are many assumptions here that are beyond the scope of this book. Not all people will behave in this way, but the general trend is clear. Furthermore, from Figure 5-4 there are morale problems already (31), which makes 105 even worse. Blame generally results in more distrust and less motivation (104), and spreads it across more levels of the workforce.

The loop between boxes 104 and 105 can be very serious. It can become a self-perpetuating cycle, with mistrust breeding criticism, which breeds more mistrust, to the extent that even eliminating everything below the loop doesn't solve the problem. In such a situation the fundamental lack of trust becomes itself a core problem. This loop can frequently be seen in union environments, where lack of respect and understanding between union and management magnifies problems to the extent that a win–win relationship becomes very difficult to achieve.

Key Concepts

- The core problems from Chapter 5 contribute to frustration, lack of motivation, and mistrust.
- Cycles of mistrust can destroy morale.

Questions for Further Thought

1. What do people need to understand in order that the cycle of mistrust in Figure 6-1 can be broken?

2. In Chapter 2, we saw that Joe had free time that Paul could have used. Joe would only be willing to help Paul if there were money available in the budget; even though this "money" remains internal to the company. Frequently, available resources are not used because of internal accounting rules. Connect the entity "Sometimes internal rules prohibit the use of available resources" to Figure 6-1. What problems does it make worse?

7 | Mirabilia

W e want people to improve their ability to produce quality products quickly. We want to overcome the inevitability of late, over-budget, and under-performance projects. To do this, we have claimed that we want to attack the core problems in boxes 1 and 3 from Chapter 5: (1) workers are measured on how busy they are, and on whether they meet their commitment dates, and (3) things go wrong, a.k.a. uncertainty, a.k.a. Murphy's Law. There is much at stake as we do this. We are challenging the inevitability of hitherto insurmountable problems. It seems impossible. What should be our approach?

Consider this statement of a basic principle of scientific deduction as expressed by the grandmaster himself, Sherlock Holmes in Sir Arthur Conan Doyle's *The Sign of Four*: "How often have I said to you that when you have eliminated the impossible, whatever remains, however improbable, must be the truth?" Do you agree with it? What about this statement from Douglas Adams' *Dirk Gently's Holistic Detective Agency*: "'The whole thing was obvious!' he exclaimed, thumping the table. 'So obvious that the only thing which prevented me from seeing the solution was the trifling fact that it was completely impossible.'"

The truth is, the impossible is much more likely than the improbable if we've made a bad assumption, and we all make bad assumptions from time to time. There are many techniques available for solving "impossible" problems.[13] In this chapter we will discuss a number of impossible things that are required before we have real solutions to the real problems.

"There's no use trying," she said: "one *can't* believe impossible things."

"I daresay you haven't had much practice," said the Queen. "When I was your age, I always did it for half-an-hour a day. Why, sometimes I've believed as many as six impossible things before breakfast."

Lewis Carroll, *Through the Looking-Glass*

We next propose some miracles that seem to be required in order to solve the problems presented so far. You don't have to believe they're possible yet; in the remaining parts of this book we will explore some of the assumptions that need to be changed. Changing a few key assumptions allows us to change the solutions from miracles to common sense. You will need to suspend your disbelief for only a short time. The required miracles that we must cause to happen are the following:

1. We have an approach to scheduling and logistics that protects us from the effects of Murphy's Law.
2. People are focused on global (system-wide) improvements rather than local ones.
3. Everyone understands and accepts the policies, procedures, and measurements that apply to them.
4. We believe we can make dramatic improvements.

Let's start with the first miracle. Suppose we have people estimate average (rather than inflated) task times, and workers don't need to meet their commitment dates. This doesn't seem so amazing. On the other hand, if people estimate real average times for their tasks, even if there is no multitasking, they will be late half the time. This is inevitable, and it also leads to late projects. That means these nonmiracles will never stand by themselves. If we really want people to give average estimates for task durations, if we really want them to stop multitasking, we'll have to do more than suggest that they change. We'll have to convince them that they're better off this way. We'll also have to convince ourselves that everyone is better off this way. We clearly need a means of attacking uncertainty.

We're attacking box 3 in Figure 5-2, Murphy's Law. We haven't said we want to eliminate uncertainty; we've said that we need a better approach to dealing with it, for the reasons given in Figures 5-2 to 5-4. Without such an approach, we may as well throw in the towel. Therefore, our first miracle is:

1. **We have an approach to scheduling and logistics that protects us from the effects of Murphy's Law.**

The inevitability of late tasks makes this hard to swallow. There is a glimmer of hope: late tasks do not necessarily imply late projects, despite our blithe assertion in Figures 5-2 to 5-4. We will address this first miracle in Part II.

We claim that late tasks don't necessarily imply late projects; it may also be that busy workers don't necessarily imply more bottom-line results for the company. Let's look at a list of what we want, and what we do to get it.

What We Want	What We Do
Make projects on time	Try to make tasks on time
Produce more projects	Try to make people more efficient
Shrink project times	Try to shrink task times
Projects are within budget	Detailed risk analysis
Customer satisfaction	Make more detailed specifications

There is a trend here. The things we want tend to be globally oriented: more projects, projects on time, satisfied customers. The things we do tend to be locally oriented: making tasks on time, making people more efficient, and so on. There is a mammoth effort involved in the kind of improvements suggested in the table above, and **we know** that they don't always help. Frequently, people work very hard to finish a task on time; they do their best; sometimes they work extra hours. Then they discover that the next person in line is not even ready for it. They are trying for local optima, which don't necessarily add up to global optimization. If Joe stays busy, it's hard to predict who will be made happy, but it's hard to make a case that keeping Joe busy would be a significant improvement for the company.

We are looking for system-wide improvements. We are looking for more, better, and faster projects. We need to make our next impossible statement:

2. **People are focused on global (system-wide) improvements rather than local ones.**

The system in this case is the organization. The truly important improvements must affect the entire organization. We shall attack Figure 5-2 from the bottom, by focusing globally rather than locally. It seems like a real miracle — both that it can work, and that people will buy into it.

This is a good time to think about how organizations work together. Sometimes we view people as carrying out individual tasks in isolation or in small groups, with management assigned the job of scheduling, supervising, and mediating to make sure global objectives are achieved. In reality, organizations are chains of dependent tasks and resources. Nothing exists in isolation. One task must be completed before another can be started; Joe must finish before Paul can start.

Because organizations are chains, they have weakest links. A problem in one area can affect many other areas. One late task can make other tasks, even other projects, late. By the same token, organizations also have non-weakest links. These are links for which an improvement does not help the organization. For example, in a project the Critical Path (or the more refined "Critical Chain" discussed in Part II) can be considered a kind of weakest link. Since it is the longest path through a project network (in terms of time), an improvement on the Critical Path implies earlier completion of the project. Globally speaking, we can also say an improvement off the Critical Path is irrelevant to completing the project earlier. In Part III we'll see ideas for addressing miracle 2, including the meaning of focusing globally and more dangers of focusing locally.

So far we've dealt somewhat abstractly with changes in people's behavior. We have asserted that workers don't have to stay busy, that the cutting of task estimates during the bidding process can be eliminated, in short that people will accept our impossible new ideas. Is this realistic?

Most people spend years fighting the same fires with the same hoses and axes. Accusing them of arson will not cheer them on. It's very well for Cassius to talk about the fault being in ourselves; he didn't believe it for a minute.[14] We need another miracle:

3. **Everyone understands and accepts the policies, procedures, and measurements that apply to them.**

We're now talking about globally oriented policies, procedures, and measurements, some of which may fly in the face of past practice. It will not be a trivial achievement. In fact, this statement seems extreme. Why everyone? Suppose the highest level of management decrees "henceforth, you need not worry about staying busy." The workers' changes in behavior will definitely not last past (pick all that apply):

- The next layoff
- The next time their honest estimate of work is ignored
- The next performance review

Does everyone need to understand everything? Definitely not. They need to understand what is relevant for them to do their jobs well. But if the rules are changing, thinking must be done about what it means for them to do their jobs well, and what they need to understand as a result.

The specifics of this miracle will inevitably depend on the environment in which it is applied. Part IV provides some general how-to's for the process of implementing the necessary changes, including both things that should be done and things that should not be done.

This leaves us with many questions and much work. The global focusing mechanisms must be defined. Policies, procedures, and measurements must be devised and evaluated based on global considerations. The task seems daunting, which makes our last miracle the most important, and often the most difficult:

4. We believe we can make dramatic improvements.

Why dramatic? Why not? Why limit ourselves? If we don't believe that dramatic improvements are possible, we won't look for them. If we don't look for them, we surely won't find them.

Still, many people have a hard time accepting even temporarily that dramatic improvements are just waiting to be found. Some have beaten their heads against the wall too many times to look for a door, or to be very pleased if they find one. Others may recognize that dramatic improvements require a different mindset, and may be frustrated at not knowing what that mindset should be. For now, we merely assert that part of the mindset requires accepting the above statement.

How can we believe that dramatic improvements can be made without knowing what we mean by an improvement? It's important to be clear on this, because this understanding is necessary for an understanding of Theory of Constraints. Does an improvement mean that more people stay busy? Does it mean that more people finish their tasks on time? The only kind of improvement that could really be considered dramatic is an improvement to the entire organization or system under discussion. We again validate the need for miracle 2.

Miracle 4 will, we hope, come to pass as you study, understand, and apply the concepts in this book and others.

Key Concepts

- The impossible can become probable if we can challenge bad assumptions.
- We need to protect projects from uncertainty.
- People should be focused on global rather than local improvements.
- We need people to accept the policies, procedures, and measurements that apply to them. In other words, they must accept the needed changes.
- We must believe we can make dramatic improvements.

Endnotes

1. Starting work on one project as another is phased out counts as multiple tasks.
2. For a useful discussion of high vs. low inventory levels in manufacturing, see Goldratt, E. M. and Fox, R. E., *The Race*, North River Press, Croton-on-Hudson, NY, 1986, 32–67.
3. See Goldratt, E. M., *What is This Thing Called Theory of Constraints and How Should It Be Implemented*, North River Press, Croton-on-Hudson, NY, 1990, 51.
4. This is a standard concept; for a brief overview (somewhat different in detail than presented here), see Project Management Institute Standards Committee, *A Guide to the Project Management Body of Knowledge*, Project Management Institute, Upper Darby, PA, 1996, 11–15.
5. For many interesting examples, see Schaffer, R. H., *The Breakthrough Strategy*, Harper & Row, New York, 1988.
6. The original law reads "Work expands so as to fill the time available for its completion." Parkinson, C. N., *Parkinson's Law*, The Riverside Press, Cambridge, 1957, 2.
7. Parkinson himself uses the effect-cause-effect approach to validate his law, by associating growth in administrative personnel with decline in actual work supervised.
8. Goldratt, E. M., *Critical Chain*, North River Press, Great Barrington, MA, 1997, 124.
9. Brooks, F. P., *The Mythical Man-Month*, Addison-Wesley, Reading, MA, 1995, 16.
10. Brooks, F. P., *The Mythical Man-Month*, Addison-Wesley, Reading, MA, 1995, 17.
11. Construction of CRTs is not inherently complex, but a detailed how-to is beyond the scope of this book. For more information, read Goldratt, E. M., *It's Not Luck*, North River Press, Great Barrington, MA, 1994; Dettmer, H. W., *Goldratt's Theory of Constraints*, ASQC Quality Press, Milwaukee, WI, 1997; Stein, R. E., *Re-Engineering the Manufacturing System: Applying the Theory of Constraints*, Marcel Dekker, New York, 1996; or Noreen, E., Smith, D. and Mackey, J. T., *The Theory of Constraints and its Implications for Management Accounting*, North River Press, Great Barrington, MA, 1995.
12. Boehm, B. W., *Software Engineering Economics*, Prentice Hall, Englewood Cliffs, NJ, 1981, 40.

13. Edward deBono's work on creativity is well known. Lev Altshuller's book *And Suddenly the Inventory Appeared*, Technical Innovation Center Inc., Worcester, 1994, presents rules for solving difficult problems. The "evaporating cloud" technique discussed later in Chapter 11 can help resolve "impossible" conflicts.

14. The full quote (emphasis added) is: "The fault, dear Brutus, is not in our stars,/ But in ourselves, **that we are underlings**." William Shakespeare, *Julius Caesar*, Act I, ii, 140–141. Cassius blamed management.

CRITICAL BUT STABLE

II

In Part II we need to accomplish the following miracle from Chapter 7:

1. **We have an approach to scheduling and logistics that protects us from the effects of Murphy's Law.**

This is extremely important; if we can't deal more effectively with uncertainty, our hope for dramatic improvements dies at once.

The term "Critical Chain scheduling" refers to a practical, common-sense method for scheduling and managing projects. By adding some simple ideas to traditional Critical Path analysis, we can produce much more reliable schedules. For the project manager, such schedules can be used as an effective means of prioritizing resource use, focusing day-to-day efforts, and reducing chaos. For the organization as a whole, this approach can dramatically cut project durations, increase reliability, and improve the bottom line. In general, Critical Chain scheduling is an effective tool to protect projects from uncertainty. It is probably the most important new development in project scheduling in more than 30 years.

8 A Critical Chain Schedule

Rather than dive immediately into a technical discussion, we'll start with a story that illustrates the ideas behind Critical Chain scheduling. The scheduling approach itself consists of the following major steps:

- **Identify the key tasks.** These are tasks on which the ultimate duration of the project depends, also known as the Critical Chain.
- **Exploit performance on the key tasks.** We do everything we can think of to make sure the key tasks aren't late, including creating a detailed schedule.
- **Subordinate to the key tasks.** We make sure that everyone else is doing their best to ensure that the schedules of the key tasks are protected.

In the following the story, we're back with our intrepid project manager Janet, this time on a weekend.

Saturday Morning Blues

I'm lying in bed, trying to plan the day ahead and at the same time grab a few more precious minutes of relaxation. It's a losing battle, but it's better than the alternative, which is getting up. It's 8 A.M. on Saturday morning and I have to run six errands around town. I hate driving through town because anything can happen. The drivers and the traffic are crazy. If everything works out, any errand might take 5 minutes; if I really get stuck, it might take 15.

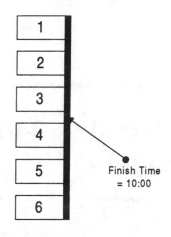

Figure 8-1   Tasks and Dependencies

Ten minutes would probably be the right average, if averages ever really happened. I sure would like to stay in bed a few more minutes, but I need to be home by 10:00 to give the car to Jack so he can drive our daughter Kerry to her karate lesson. I wonder when I should leave home? Oh, I almost forgot — my third errand is to stop by Emily's house to pick up a bagel slicer. Her schedule is tight, too, so I'll need to phone and let her know when I'll be by. I wonder what should I tell her?

Interlude: Schedule #1

We have two questions to answer: when should Janet leave, and what should she tell Emily. First, it's useful to lay out the tasks and task dependencies, as we've done in Figure 8-1.

The boxes represent the different tasks that Janet must complete by 10:00, when Jack and Kerry must leave. Task durations are proportional to their horizontal length, in this case 10 minutes each. The vertical bar indicates the end of the "project." We can work backwards from the finish time, placing the boxes one after the other like a row of bricks, to find the required start time. If we stacked the tasks backwards in this way from 10:00 we would reach a start time of 9:00. Since 10 minutes per task is only the average, the chances are only 50–50 that she could finish everything in an hour; she really should leave some extra time.

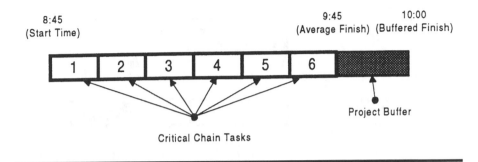

Figure 8-2 First Schedule

The worst case is 15 minutes per task. If Janet wanted to be absolutely certain of making it home by 10:00, she would increase the task durations to 15 minutes each and leave by 8:30. Realistically, the chances of everything going wrong are slim, and she could no doubt leave later than that. How about 8:45? We hope that the good fluctuations — the tasks that take less than 10 minutes — will help cancel out the tasks with bad fluctuations that take more than 10 minutes. We expect she would get home before 10:00, but we've left some protection: a "buffer" of 15 minutes. This protection is called a **project buffer**. It's there to protect the completion time of her "project."

The set of tasks that determine when a project can finish is called the **Critical Chain**. They're critical because an improvement anywhere on the Critical Chain means the project can get done earlier. The faster your Critical Chain tasks go, the sooner you can finish the project. They're a chain, rather than a path, because they take into account resource dependencies. Clearly Janet cannot do all her errands simultaneously; they must be spread out over time, as shown in Figure 8-2.

When blocks come horizontally one after the other, the tasks must be performed sequentially. For example, the task labeled "3" must come after those labeled "1" and "2". The gray block represents the project buffer. We've made that buffer an explicit block because, as we'll see in Chapter 9, it's a key component of a realistic schedule. It is not slack time. Critical Chain tasks have been given thick borders which will set them apart from other tasks. (In this example, all tasks have thick borders because all are part of the Critical Chain.) Notice that the project completion time will be affected by fluctuations in any of the six Critical Chain tasks.

Now let's think about when Janet is likely to arrive at Emily's house. If she starts out at 8:45, and we assume travel to Emily's house is the main part

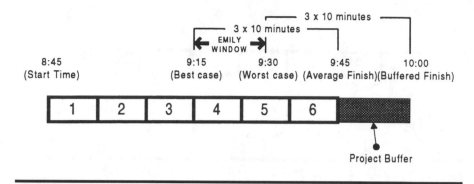

Figure 8-3   First Schedule (Intermediate Due Date)

of errand 3, she would meet Emily somewhere between 8:45 + (3 × 5 minutes) = 9:00 (the best case), and 8:45 + (3 × 15 minutes) = 9:30, the worst case. In this case it's much more accurate not to be precise, but rather to specify a window of time during which Janet expects to arrive. She shouldn't say "I'll be there at 9:15"; chances are that will be wrong.

Suppose Emily needs a smaller window than a half hour; what then? Janet could arbitrarily say "9:10 to 9:20." But if she had bad luck at the beginning, she might not quite make it by 9:20. Suppose instead we back off from 10:00 by our 15-minute buffer, to 9:45; then go back by the average task duration — 10 minutes — to "schedule" each task. That puts the meeting with Emily, the finish of Task 3, at 9:15, as in Figure 8-2. If Task 3 took place at 9:15, Janet would still have 10 minutes each for the three remaining tasks, plus the full 15-minute buffer. So even if she arrived at Emily's early, she knows she can wait until 9:15 and still have the full buffer as protection. Of course, things might go poorly with the first tasks; she might use up some of the buffer and arrive after 9:15. How much after 9:15? Certainly no more than a full 15-minute buffer. So it looks like Janet would be safe specifying a window of 9:15 to 9:30 for a time to meet. Figure 8-3 shows the schedule with this window defined.

The "best-case" finish time for all Janet's tasks is at 9:45. Going back from 9:45 by the three 10-minute tasks gives the best-case start time for the stop at Emily's, 9:15. The "worst-case" finish time for Janet's tasks is 10:00; going back from there by three 10-minute tasks gives a worst-case time of 9:30 for the stop at Emily's. The window of time expected for the finish of Task 3 is therefore 9:15 to 9:30.

Imagine for a moment that each of these errands required that Janet arrive at a specific time — everyone else has a precise schedule to keep, too. If she scheduled them at 10-minute intervals, she'd be virtually certain to be late for some. The positive fluctuations could not accumulate, unlike the negative fluctuations. To avoid being late she'd have to schedule the errands every 15 minutes. This would mean putting in a buffer of 5 minutes for each task, which means we're back to the worst case of leaving at 8:30.

By putting a project buffer at the end of the set of tasks, you can protect all the tasks. By specifying windows of time rather than precise times for intermediate tasks, you allow yourself to capitalize on any good fluctuations that might happen by starting tasks earlier than expected. If you have to buffer each task, you end up spending too much time protecting your promises. These **intermediate due dates** can have a significant cost.

If Janet could speed up tasks on the Critical Chain, she could give herself time to do other important things, such as reading the newspaper. She will probably look for better routes for her errands, think about where to park, and generally try to get through things as quickly as possible. The one thing she shouldn't do is drive recklessly. Careless actions early in the project can have magnified effects later on.

The Blues, Verse 2

Lying in bed thinking about it, everything seemed too easy. It was. Somebody hid my keys and now I'm running late. It's 9:00 and I have to leave. I know I have at best an even chance of making it back by 10:00. That means when I arrive home, my daughter Kerry had better have her clothes on and be ready to go. How can I be sure that will happen?

Interlude: Schedule #2

She can't, obviously. Janet has an hour in which to do an hour of work; her daughter has an hour to do 5 minutes of work. It seems certain that without a major cataclysm they won't be able to leave until at least 5 minutes after Janet gets back.[1] Kerry's getting dressed isn't part of the Critical Chain, because whether she's ready at 9:30 or 10:00 won't matter. But she still has the power to make everything late. Ideally, she should start getting dressed

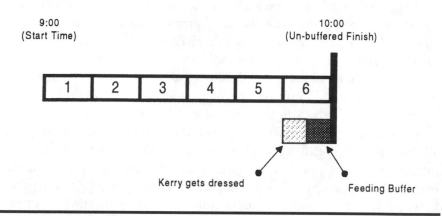

Figure 8-4   Second Schedule

10 minutes before Janet gets back. That not only leaves time for her to get dressed, but also some slop. That slop might be useful in case she can't find the right clothes; it might also be useful for her to be ready in case Janet arrives home early. This "slop" is called a **feeding buffer**: her task (getting dressed) "feeds into" Janet's Critical Chain tasks (the errands). Since any delay of Janet will delay the whole schedule, she doesn't care when Kerry starts getting dressed, but she definitely wants Kerry to be ready by the time she gets back. It's probably better if Kerry doesn't get dressed too early, since she might get her white robe dirty, so Janet quickly asks Jack to make sure Kerry starts getting dressed by 9:50. Then she makes him repeat the instruction. Maybe Janet has a chance (Figure 8-4). Notice the vertical bar, which again signals the completion of this "project." The finish time is marked as "un-buffered," which indicates that there is no project buffer.

The Blues, Verse 3

Oh, one more thing…. While I'm away my husband will be trying to relax, as usual on a Saturday morning. But if he wants to be dressed before he goes, he'll really need to do some laundry, which will take about 40 minutes. The karate lesson is at 10:30. It's about a 20-minute drive to get there, but sometimes there are delays; besides, it doesn't hurt to be early. I wonder how things fit together now?

Interlude: Schedule #3

We've extended Janet's scheduling job to include Jack's task, as well as the trip to karate class. Now we want to insert the project buffer ending at 10:30, and (like it or not) find we have only 10 minutes for it. The Critical Chain consists of Janet's errands plus the trip to karate. Any improvement along this path will speed everything up; any delay jeopardizes the 10:30 lesson time. Those tasks have to be scheduled sequentially. Getting Kerry dressed, including a 5-minute feeding buffer, still means a 9:50 start for that task. The remaining question is where to schedule the laundry.

The feeding buffer concept still applies. That is, if you scheduled laundry to begin at 9:20 there would likely be mishaps that would delay the trip to karate. If there is a buffer of 10 minutes between laundry and the trip to karate, Jack should have a good chance of being ready for the trip and protecting the Critical Chain schedule.

We also put in another type of buffer, for both Kerry and Jack, to make sure they don't wander off and start something else. This is called a **resource buffer**, because it is put in to make sure the resources (Jack and Kerry) are ready to start on their Critical Chain tasks. They have to be ready to roll when the time comes, and this buffer will help alert them that it's time to start.

The final schedule is shown in Figure 8-5, with Jack's task labeled "Laundry". The thick arrows represent task dependencies, just as with adjacent boxes. That is, "6," "Laundry" and "Kerry gets dressed" must be complete before the "Trip to Karate" can start. The "Trip to Karate" requires both Jack and Kerry to be available for the trip. We've also put in resource buffers as wakeup calls for both Jack and Kerry, so that they are ready to start the trip. Not only must Jack and Kerry finish their tasks before the trip: they themselves must also be available to make the trip. Their time must be reserved, which requires a resource buffer. For Kerry, both the task (getting dressed) must be finished, and the resource (Kerry) must be available. We'll see more examples of multiple buffer types in the same place in Chapter 12.

Notice that the resource and feeding buffers don't increase the length of the Critical Chain. They act like the shock absorbers in a car, protecting it — and hence the completion of the "project" — from disruptions. The project buffer protects another important part of the system, the commitment date, from disruptions on the Critical Chain. Notice also that resource and feeding buffers don't just protect the Critical Chain from problems (negative fluctuations) on non-Critical Chain tasks. They also allow the chance for Critical

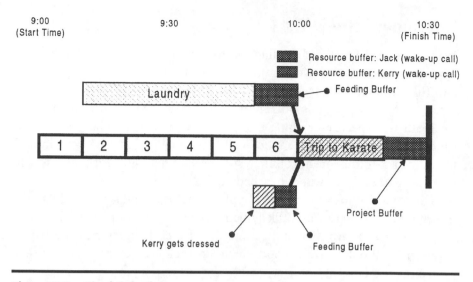

Figure 8-5   Final Schedule

Chain tasks to start early, in case those tasks go better than expected (positive fluctuations).

What are the chances of any schedule, including this one, being followed in reality? Pretty slim. But the overall objective isn't really to "meet the schedule." Taking a more global viewpoint, the family (our system) needs to meet their commitment, which is the 10:30 lesson time, while carrying out all the intermediate tasks, and not getting in a car accident in the process. We've improved the chances of success, both by creating a realistic schedule that takes into account the important requirements, and by placing strategic buffers. That's the technical part. Even more important, we've raised everyone's awareness of what is needed in order to work together effectively. We followed the steps presented at the beginning of this chapter:

- **We identified the Critical Chain.** Now everyone knows how their tasks relate to the most critical tasks, and hence to the overall objectives.
- **We exploited the Critical Chain.** That is, we did everything we could think of to improve performance on Critical Chain tasks, including creation of a detailed schedule with no contention for resources.
- **We subordinated to the Critical Chain.** We identified what everyone needs to do to support the Critical Chain schedule. We put in buffers to help protect the Critical Chain tasks. We can measure the impact of following, not following, or improving on the schedule by checking the impact on the Critical Chain.

These points correspond directly to Steps 1 to 3 of the five-step process that we will discuss in Chapter 17. In addition, by identifying the Critical Chain we have discovered where to apply thought and resources to "crash" the project, should it be necessary to shrink the project duration further. We could bring in people from other projects, hire new people, hire consultants, buy equipment or software, or assign overtime. Janet might get Jack to run an errand later. If you try to add resources without identifying the Critical Chain, you may be wasting them. This is Step 4 of the five-step process, "Elevate."

The scope of the schedule is very important when key resources are involved. Imagine for a moment that Jack's schedule late in the day is very full. It's possible that he has later appointments that must be kept. In order to keep these appointments, some of the other tasks, such as taking their daughter to her lesson, might need to be scheduled even earlier. Perhaps Janet would need to get back in time to make sure Kerry gets dressed. In this case the makeup of the Critical Chain would change; most of the tasks on it would be done by Jack. Janet's lost keys might have lost Jack the chance of meeting his commitments later in the day. If possible, it's best to examine all the project's tasks that can affect each other significantly, whether the dependencies are due to task sequencing or shared resources.

A practical person might be thinking "why didn't they just start earlier?" We often ask that question during the later stages of projects. Of course, sometimes it's not possible to start earlier. Sometimes there is a real conflict between starting earlier and starting later. On the one hand, starting earlier can create more work-in-process, resulting in more confusion and the potential for more rework, if for example, specifications change. In addition, it's usually best to postpone cash outlays for investment as long as possible. On the other hand, starting too late could have an impact on the project completion date.

Many compromises have been proposed to this conflict. The Critical Chain method is not a compromise. Tasks are started when they need to be started, based on the buffer protection required. They are not started earlier, because that is unnecessary; they are not started later, because that could jeopardize the project completion date.

Perhaps Janet frequently loses her keys, in which case the problem may be what is known as a **policy constraint**. Erroneous, ill-considered, or out-of-date policies can adversely affect an entire system. We'll look at some common policy constraints in Part III. First we must develop the Critical Chain concept in more detail.

Key Concepts

- The **Critical Chain** is that set of tasks which determines overall project duration, taking into account both precedence and resource dependencies.
- Improvements along the Critical Chain will likely result in improvements to the project as a whole; improvements elsewhere will not.
- The **project buffer** protects project commitment dates from fluctuations on the Critical Chain.
- The **feeding buffer** protects the Critical Chain from fluctuations on feeding tasks, and provides the possibility for Critical Chain tasks to start early.
- The **resource buffer** protects the Critical Chain from lack of availability of required resources, and also provides the possibility for Critical Chain tasks to start early.
- In general, **intermediate due dates** increase the overall project duration; instead, examine progress by looking at the buffers.

Questions for Further Thought

1. How would the buffered schedule in Figure 8-5 look if Jack had two 40-minute tasks that had to be completed before he could drive Kerry to her lesson? Assume both Jack and Janet can start before 9:00.
2. Would the schedule in Figure 8-5 have made any sense if Janet had used worst-case task times rather than averages?
3. Which task completion times did Janet care about from the schedule in Figure 8-5?

9 Managing Uncertainty

f buffers were just a planning tool, a means of getting shorter and more reliable schedules, that would be sufficient reason to use them. But there's much more. Buffers are an extremely valuable tool for monitoring the status of projects and determining whether drastic actions are required. In this chapter we will look at these aspects of buffers.

Managing Janet's Buffers

Let's look back at Figure 8-5 and consider the effects of some disruptions to Janet's schedule. Suppose Task 1 takes 15 minutes, instead of the average 10. Janet has gone past the expected finish time by 5 minutes. The project buffer being fed is 10 minutes long; she has eaten into it by 5 minutes; there are 5 minutes of buffer time remaining. If Task 2 takes 15 minutes as well, the entire buffer is eaten up. Any further delays stand a good chance of jeopardizing on-time arrival at the karate lesson. Janet may have to consider drastic measures, such as postponing errands or letting someone else do them.

As a good manager, Janet decides to monitor the status of her other buffers as well. While in the midst of her errands, she picks up her car phone at 9:30 and gives Jack a call. "How are your jobs going?" she asks. He says he's 7 minutes behind, but should make the 10:00 time. "Don't forget to get our daughter dressed at 9:50," she says.

At 9:52 she checks the feeding buffer by calling and asking about their daughter. "Oh, I forgot!" Jack says. He hasn't started getting her dressed. Fortunately there's still time. And since Janet is running late rather than early, there's no chance of an early start for the karate trip, so the problem is solved by the call. She tells him that she expects to arrive at 10:05, and says they should be ready by then. In that way she applies the "resource" buffer to make sure Jack and Kerry are ready to leave.

65

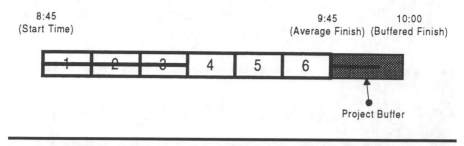

Figure 9-1   Buffer Usage

In order to determine the effects of different delays on her "project," Janet didn't have to wait until the Critical Chain tasks or the final completion time were delayed. She could measure the impact immediately by looking at the buffer. In fact, on all paths leading to a buffer it's useful to monitor the following data elements:

- Current active task
- Amount of buffer remaining unconsumed
- Duration of the chain of remaining tasks feeding the buffer

This gives you an idea of how much buffer has been consumed, vs. how much of the processing has been completed. If you're at the start of a path and the entire buffer has been consumed, you have a problem. If you're at the end and no buffer has been consumed, you'll be early.

Things are not always so clear. Let's look back at the simple example of Figure 8-2, and imagine that half of Janet's errands have been completed. However, the delays on those tasks have cause two-thirds of the buffer (about 10 minutes) to be used up. The picture is shown in Figure 9-1.

The horizontal lines through tasks represent completed tasks; the horizontal line partway through the buffer indicates that it is partly used up. How do you evaluate this situation? There may be a problem. Probably Janet should start considering other options. However, much depends on where the high-risk tasks are in the path. If the high-risk tasks have already been completed, if (for example) there isn't much traffic in the neighborhood of the remaining tasks, Janet is doing well. If there is a lot of risk still to be encountered, this buffer may not be enough protection.

There are various ways of quantifying risk, such as looking at "average" vs. "worst-case" task durations. This kind of explicit quantification could be included in the buffer analysis, but usually the available data aren't accurate enough to warrant it. The buffers can be checked "by eye," which in most cases will be good enough. Project managers usually have sufficient experience and intuition to judge the amount of risk remaining, and whether the remaining buffer is sufficient. Buffers must be checked.

Risk Placement

This discussion suggests a useful rule: **Schedule high-risk tasks as early as possible.** If there is a problem, buffer management will allow you to detect it early. You also have more options for solving it. If the high-risk tasks are scheduled late, there may be no room to solve problems.

Let's consider, for example, a project that has formal acceptance test procedures. The tests must be passed before the project can be signed off on. Some of the tests are trivial and some are extremely rigorous. It's very possible that the acceptance tests will result in changes, which in turn require fixes and then re-running the tests: loops of rework. This is often considered a part of doing business when developing (for example) military hardware. The problem is that high-risk tests are often left as part of formal acceptance, which occurs toward the end of the development process. This frequently results in delays that make projects late or later.

Suppose we split the testing process apart from the formal acceptance process. As a module is completed, the rigorous testing is done as early as possible early to discover potential problems. The formal parts of the process, which include (for example) sealing up the unit and doing less important tests, don't have to be done at that point. What we want is to minimize the impact of surprises later on when there are fewer options to deal with the delays they will cause.

Buffer Types

So far we have seen three types of buffers. The most important is the project buffer. It is placed at the end of the project, in order to protect the customer commitment date from disruptions along the Critical Chain. It is extremely important to monitor the project buffer, in order to identify global problems before they become serious.

The feeding buffer makes sure that work needed for the Critical Chain is available from non-Critical Chain paths when it is needed. In this way it protects the Critical Chain, and thus indirectly the customer commitment date. The feeding buffer is placed between Critical Chain tasks, and the paths that join them.

While feeding buffers make sure that the work is available, resource buffers make sure that the resources are available to do the work.

But there are a number of questions that arise when we consider treating resource buffers in the same way as feeding buffers:

- Should we actually reserve space on resources? In other words, should we schedule resource buffers as "idle" time on the resources, or should we just make sure that the Critical Chain tasks have priority when they are ready to begin?
- What is the real capacity of a resource, given that people can work overtime (often with no pay)?
- How big should the buffers be? In other words, how can we quantify the risk we're protecting against?

In general the resource buffer can be treated as just a wake-up call, which alerts resources to be ready to work on their Critical Chain tasks when needed. This is the simplest solution, and for most purposes good enough. For example, 2 weeks before the Critical Chain task needs to begin, the resource (or perhaps the resource manager) is notified that they'll be needed in 2 weeks. They are again notified 1 week before, and then 2 days before. This gives them plenty of time to make sure they can be ready. It also keeps them apprised of changes to the Critical Chain schedule.

If the resource buffer is treated as a wake-up call, we can make an additional recommendation: if there seems to be a significant risk that a resource will not be available to start a Critical Chain task, push that resource's tasks earlier, when they feed the Critical Chain task, in order to leave space for a resource buffer.

Rescheduling

When is it necessary to reschedule? Traditionally, rescheduling is done frequently, perhaps every week. There are a number of ramifications of frequent rescheduling:

- Priorities of workers and managers are constantly changing.
- Changing schedules result in mixups and confusion.

- The Critical Chain changes, and thus the points that are important to protect change.
- The buffer locations change, and hence our monitoring efforts change from week to week.

This would imply that schedules should change as infrequently as possible. Of course, schedules must be monitored, and we will have to take care in the following situations:

- New projects are introduced, or old ones canceled: we must take existing resources into account.
- Non-Critical Chain tasks go more slowly than expected: we monitor the feeding buffers.
- Critical Chain tasks go more slowly than expected: if the project buffer proves insufficient, drastic actions might be needed, including informing the customer of a delay. It's almost always better to inform customers in advance (using the buffers as a guide) than after the delay is a fact.
- Non-Critical Chain tasks go more quickly than expected: there might be available resources which can be used for other tasks or projects.

For all of these cases we want to monitor the schedules. Seldom do we want to create new schedules; almost never do we want to change the composition of the Critical Chain.

When is it necessary to communicate changes in the plan to workers? There are a number of obvious times, such as when tasks are eliminated or added, or where there are changes to the composition of tasks. Changes should also be communicated when the level of intensity required of someone changes. Required intensity level should change when buffers are used up more quickly than anticipated.

Most changes to plans do not need to be communicated to workers. When priorities or needed level of effort don't change, there is seldom a need to communicate changes. In Chapter 21 we will discuss the specifics of what various people need in their schedules.

Information Flow

Without the buffers, there is frequently no good way to tell when a late task is a serious problem. Consequently a project manager is likely to under- or

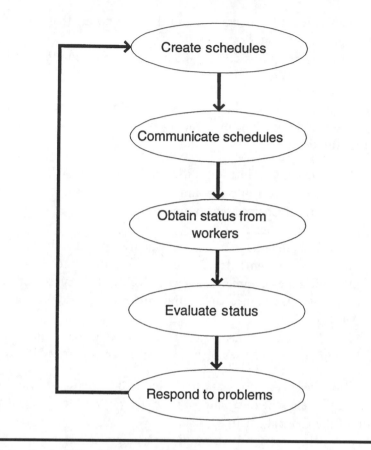

Figure 9-2   Project Tracking (traditional)

overreact. The project manager also has no good way of justifying a feeling that perhaps he or she needs the resources that are being moved elsewhere. In order to know what's going on, the project manager must revise the schedules. The typical flow of events is shown in Figure 9-2.

Schedules are created and then communicated; status is obtained and evaluated; problems are responded to, and rescheduling occurs. The errors and lost productivity caused by reprioritization can result in an unfortunate feedback loop: more problems implies more reprioritization, which itself generates more problems. Now consider the Critical Chain approach (Figure 9-3).

Now the loop occurs between obtaining status, evaluating problems, and responding to serious problems. This approach allows a measured response to problems. The impact of delays can be assessed by looking at buffers. If the delays don't have a significant impact on the project as a whole, they will

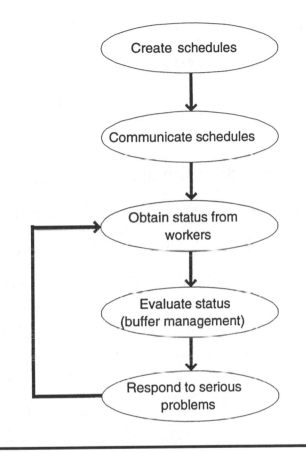

Figure 9-3   Project Tracking (buffer management)

not require significant corrective actions. The need for frequent rescheduling, and the corresponding confusion, is cut dramatically. Management actions to reshuffle resources can be evaluated in terms of the impact on the buffers, and hence the impact on the projects. Decisions can be evaluated much more rigorously.

Key Concepts

- Buffers protect project completion by protecting key areas of the schedule.
- Buffers can be used to look ahead and predict the effects of schedule disruptions on the project as a whole. This ability to look ahead gives more time to fix problems.

- Using buffers, managers can have more confidence in their decisions and in their ability to justify decisions. This means they can be more productive. You don't need to walk as slowly when the lights are on.
- Schedule high-risk tasks as early as possible.
- Buffers protect the project completion date and the Critical Chain.
- Use buffer management and avoid frequent rescheduling.

Questions for Further Thought

1. Why are shorter buffers advantageous? What kinds of actions does this imply?
2. Why are longer buffers advantageous? What kinds of actions does this imply?
3. Historical buffer data can track which resources caused buffer time to be used, and which resources caused buffer time to be made up. Think of some ways in which such data could be useful to track.

10 Resolving the WIP Conflict

In the last two chapters we looked at buffers as a means of protecting critical schedule points. These schedule points are tasks on the Critical Chain, and project completion. In Chapters 2 and 3 we discussed a major conflict experienced by workers and project managers, between increasing work-in-process to make sure tasks complete on time and people remain productive, and decreasing work-in-process to have short lead times. The buffers provide the solution. Let's look at the problem and solution in detail.

As we've seen, in the real world some attempt is often made to add safety to each task, so that the start time of each subsequent task can be maintained. This isn't usually realistic, because customers won't stand for the delivery times that result. On the other hand, if average task durations are specified everywhere and no buffers are put in place, it's guaranteed that projects will be late. It seems like a catch-22: if you pad tasks, you don't get work, and short-term profit suffers; if you don't pad tasks, projects will be late and long-term profit suffers.

There is a rigorous means of expressing any conflict, called a "Conflict Diagram" or an "Evaporating Cloud."[2] This WIP conflict is shown in Figure 10-1.

The box labeled "A" is our objective: maximize profitability. There are two requirements for this, in boxes B and C: quote shorter lead times, because it allows us to sell more projects, possibly at higher prices; and give reliable commitment dates, in order to keep the existing customers happy. These requirements in turn each have prerequisites in boxes D and D'. The arrows between A and B, A and C, B and D, and C and D' represent necessary conditions: in order to have A, we must have B; in order to have B, we must

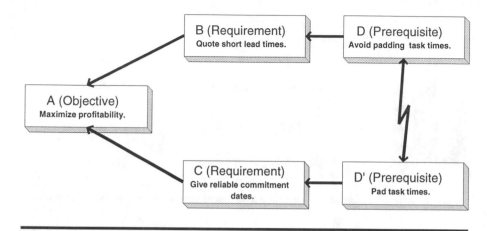

Figure 10-1   WIP Conflict

have D; and so on. We read A to B as "In order to maximize profitability, we must quote short lead times." B to D would read "In order to quote short lead times, we must avoid padding task times." Try reading all the straight arrows this way.

The conflict is between the prerequisites, boxes D and D′, and is represented by the crooked arrow. The conflict can be read "We can't both ⟨box D⟩ and ⟨box D′⟩." We can't both avoid padding task times and pad task times.

Conflicts of this type are typically solved by compromise. One compromise approach is to put in some safety — less than enough, but more than can really be afforded. This combines poor bids (suboptimal current profitability) with poor commitment performance (suboptimal future profitability). Another approach is to put in no safety, but give people an "incentive" to use their personal time as a buffer. That is, it is expected that projects will have "crunch times" during which people will work extreme hours. Sometimes these crunch times last for months. From a long-term company perspective this often leads to burnout and eventually the loss of valuable assets, namely employees, which can lead to lower current or future profitability. Furthermore, overworked people are often less productive and less creative. Many companies don't look that far ahead.

This picture can be used to derive possible solutions. Each arrow in the diagram contains underlying assumptions; if any of these assumptions can be invalidated, the arrow goes away, and hence the conflict goes away or "evaporates." Assumptions on the straight arrows can be exposed by reading "In order to ⟨box 1⟩ we must have ⟨box 2⟩ because ⟨assumption⟩." For example, "In order to maximize profitability, we must give reliable commitment

Table 10-1   Assumptions in Figure 10-1

Arrow	Assumption
A to B	Short quoted lead times are important for customers.
B to D	Safety increases lead times significantly.
A to C	Customers care about commitment dates.
A to C	Profitability depends on customer satisfaction.
C to D′	There are statistical fluctuations and unanticipated problems.
C to D′	We must deal with uncertainty by padding task times.
D to D′	All tasks need safety time.

dates, because … profitability depends on customer satisfaction." Assumptions on the conflict arrow can be exposed similarly by reading "We can't both ⟨box D⟩ and ⟨box D′⟩ because ⟨assumption⟩." Some assumptions in this diagram are shown in Table 10-1.

If any of these assumptions can be invalidated, the conflict no longer exists. Suddenly there are many directions from which the problem can be attacked. The traditional approach is compromise coupled with ignoring reality. For example, we ignore statistical fluctuations and unanticipated problems; we compromise about task times; we ignore the certainty that projects will be late.

From what we know already about buffers, we can quickly recognize that C to D′ is a good place to attack. We've seen from the last chapter that some tasks are very important to protect and some are not. All tasks are important to complete; not all are important to complete as quickly as possible. We can buffer the few key areas to protect their performance to schedule, and leave the rest alone. We can put in aggregate buffers rather than per-task padding. It becomes obvious that the conflict has evaporated if we rewrite it as in Figure 10-2. In Figure 10-2, the rewording has caused the conflict to go away; there is no longer a conflict arrow between D and D′.

At a conceptual level, we've now solved the problem posed in Chapter 3. The purpose of work-in-process is to maintain production; that is, real, bottom-line production, in this case completed projects. Otherwise, work-in-process is unnecessary or even wasteful, and should be eliminated. Buffers to protect customers and the Critical Chain are good; padding to protect noncritical points is bad. Furthermore, we know where to place the buffers. We place feeding buffers to make sure the Critical Chain is supplied with material on which to work. We place resource buffers to make sure the Critical Chain tasks have the resources they need to keep going. We place project

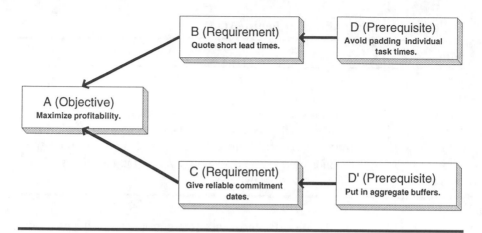

Figure 10-2 WIP Conflict Resolution

buffers in order to protect the customer from problems on the Critical Chain itself.

When we pool our protection in the places it's needed, the overall buffering for projects can be much less than it would be if the protection were spread out haphazardly. We can plan for average completion times for tasks and let the buffers take care of the statistical fluctuations. However, this also means that buffers are not optional. They are not float or slack time or lag time or time reserve. They are a key piece of the schedule. This is very difficult for people to accept if they don't understand what a buffer is. If buffers are treated as slack, they will be used up. In order to apply the buffer concept effectively, everyone must understand the reasons buffers are important.

For example, a typical upper-level manager might look at a project schedule, see the project buffer, and say "remove that slack time." Then they're likely to ask when the project can be completed. The problem is, it's highly unlikely that the project will finish before the buffer starts, because we've used average task durations. On the other hand, it's highly likely the project will complete before the end of the project buffer. Due to the inevitable uncertainty, any estimation of when the project will finish must be accompanied by a statement of probability.

It can also be important for customers to understand the meaning of buffers. If a customer sees some "idle" time in the schedule, depending on the contract and the relationship they may complain that they don't want to pay for idle time or for "planning to fail." They may want the project delivered before the end of the project buffer, since it appears to be slack. There are two ways this can be dealt with and still maintain appropriate buffers. First, you can maintain separate internal and external schedules. The customer only sees the external schedule, which has the buffer distributed among tasks. The project is managed according to the internal schedule. Second, you can educate the customer. Both approaches have risks and costs, and the choice will depend on the circumstances.

Sometimes customers want milestones. They want to know that a particular task will finish by a particular date. There are again two options. You can give the customer a date that is the expected task finish plus some buffer. The customer doesn't even need to know that the buffer is there. The second option is to educate them, so that they don't need milestones, so that they understand the meaning of buffers and buffer use.

Besides the time saved in overall project duration, buffers provide a significant additional benefit. The key points in the schedule are protected from disruptions, meaning that disruptions will usually have less global impact. Any disruption less than a catastrophe will only be propagated through the schedule until the buffer is reached; the buffers will absorb the disruptions. The standard domino effect is stopped. This in turn means that there is much less need for frequent rescheduling. The Critical Path approach requires constant monitoring and regeneration of schedules. The schedule regeneration process itself results in errors and miscommunication. The Critical Chain approach significantly reduces the need for this. We will discuss this topic in more detail in Chapter 21.

In the next chapter we move on to the mechanics of Critical Chain scheduling.

Key Concepts

- Any conflict can be diagrammed explicitly.
- There are hidden assumptions behind any conflict that can be challenged to make the conflict "evaporate."
- The conflict between less and more work-in-process can be evaporated through the use of buffers.
- Buffers are not optional.

Questions for Further Thought

1. Think of some more assumptions behind the arrows in Figure 10-1.
2. Find other ways of "evaporating" the conflict of Figure 10-1. Do they seem satisfactory?
3. Various kinds of shock absorbers in cars protect key places from fluctuations (vibrations). Think of some other situations in which protection can usefully be aggregated, rather than spread to many places.

11 Identifying the Critical Chain

I n order to use Critical Chain scheduling as more than an intuitive means of making better schedules, we'll have to be more specific about both what we want from it, and the technical aspects of creating a schedule. Ultimately, we would like schedules to have the following characteristics:

- They must be realistic, based on what we do know
- They must not require frequent changes
- They must be reliable, coping with uncertainty but as short as possible in duration
- They must show us which tasks are key, and
- They must be based on global rather than local optima.

If we achieve these prerequisites, we will have accomplished Miracle 1, an approach to scheduling and logistics that protects us from the effects of Murphy's Law.

We'll achieve better realism through use of the Critical Chain, rather than the Critical Path, because we've taken into account resource use. By specifying average durations, we'll decrease overall project duration. By using buffers, we'll increase commitment date reliability and reduce the need for frequent schedule changes. We'll relate decisions to global optima by using globally oriented measurements, and ultimately by scheduling multiple projects.

We're going to use two measurements in deciding how good a schedule is. The first is how quickly it can finish. The more quickly a project can finish, the sooner people can get on to the next project, and (broadly speaking) the more money the company will make. The example in this chapter will be for a single project. In the real world we may have to take into account some interdependencies between projects, or we run the risk of individual project performance being a sort of local optimum. We will deal with this in Chapters 18 and 19.

The second measurement is **work-in-process** (WIP). We'd like all the tasks to be scheduled as late as possible, as long as our project is still protected. Scheduling tasks later decreases work-in-process; it decreases the chances of rework if design problems are discovered; and in some cases it maximizes cash by pushing out investment until it's absolutely needed.

There are tradeoffs between these measurements, which will be discussed along with buffer sizing in Chapter 12.

In this chapter we go through the detailed process of determining the Critical Chain. We will ignore the effects of uncertainty; buffers will deal with that, and they will be placed in Chapter 12. We will construct a schedule that's realistic, if we ignore uncertainty, and that has the minimum possible WIP. First we need the initial plan, including tasks, task dependencies, and resource requirements. In order to minimize WIP, we place tasks at their late start times. We next make the plan "realistic" (still ignoring uncertainty) by removing resource contention. This process is called "load leveling" — we even out or "level" the loading of the resources. Finally, we use this schedule to determine the Critical Chain.

Step 1: Create the Initial Plan

Let's go back to the wideband receiver project that Janet bid for in Chapter 1, and see how Critical Chain scheduling might work in a more complicated environment than that of Chapter 8. This example is still simpler than most real cases will be; for example, it would be hard to justify $10 million for this project, DOD procurement procedures notwithstanding. However, it illustrates the basic technical points of Critical Chain scheduling.

As our starting point for determining the Critical Chain, we'll take a standard Critical Path layout, shown in Figure 11-1.

Figure 11-1 is the initial cut at a realistic plan. It is laid out in activity-on-node fashion, with each node or block representing a task. The horizontal axis is time, and the length of each block is proportional to the duration of the associated task. The time scale appears at the top, one month per tick mark. The picture is a kind of Gantt chart, although it is organized by the flow of work rather than work on resources.

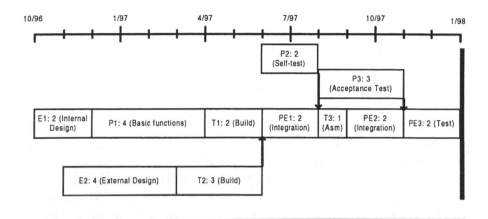

Figure 11-1 Initial Plan

There are three main resources for the project: an engineer, a programmer, and a technician. The text inside each box shows which resource (E, P, or T) the task will be performed by, a sequence number to help identify the task, and a short description of the task. For example, E1 is Task 1 for the engineer, and the description is "Internal Design." Note that tasks PE1, PE2, and PE3 each require both a programmer and an engineer.

Precedence dependencies (tasks that must be performed in a particular time sequence) are shown either with adjacent blocks or arrows. For example, P1 must come after E1; PE3 must come after both PE2 and P3. However, P2 and P3 are not on the same horizontal line or joined by an arrow, so P3 does not have to come after P2. The precedence dependencies in Figure 11-1 are all "real." This is not always the case with project plans; sometimes, for example, artificial precedence dependencies are added to make sure resources aren't overloaded.

The project itself consists of the development of an internal module and an external module. The internal module is the radar receiver, the external module its case. The internal module is first designed, then programmed, then put together by a technician. After that it's assembled with the external module (which is first engineered, then built). A programmer and engineer must then check it out. There are also self-test and acceptance test programs that must be created and integrated.

There are a few nonstandard things to note about this example. First, all tasks end as close as possible to their "late completion" times, taking task precedence into account. Tasks cannot be pushed later without butting into other tasks that depend on them. In other words, no tasks have slack or float.[3]

Second, we're using average (rather than padded) estimates of task durations. On average, if resources were dedicated solely to these tasks (no multitasking), they would be completed in the time specified. Of course, the averages won't happen in reality; that's the purpose of buffers. But it's important to note that we're building into the schedules the concept "no multitasking."

It's easy to see from the diagram that the critical path follows the tasks E1 → P1 → T1 → PE1 → T3 → PE2 → PE3. This is the longest path through the network. It's hardly realistic as a schedule, since a number of resources are overcommitted. One approach is to level the load so that resource loading is realistic, then inflate all the task times to add safety so that everything comes out on target. We don't want to do that, because we would once again spread the protection everywhere, rather than pooling it into buffers. Another popular approach is just to inflate the task times and let resource contention take care of itself; this has similar drawbacks. We want to find the Critical Chain, the set of tasks on which the duration of the project depends. To determine that, we must first level the load.

Step 2: Load Leveling

In order to level the loading of the resources, we will first go from right to left on Figure 11-1, placing the tasks on the resources as blocks of load. We will move tasks earlier if needed to make sure that resources aren't overloaded. We go from right to left, that is from future to past, so that work-in-process is still minimized by placing tasks as late as possible. We'd like to push any overloads on a given day to the past to keep the completion date as early as possible. As we start placing blocks, the placement of PE3 is easy, because there's no question about it: it must go last.

When we now try to decide between placing P3 or PE2, there's a conflict for the programmer's time because of a **resource dependency**. There's only one programmer, so we can't schedule P3 and PE2 at the same time. In order to look at this more easily, it's useful to add a new way of looking at the project. The view we've seen in Figure 11-1 is called the *precedence view;* it gives a clear picture of the precedence dependencies in the project. The new view is the *resource view,* which shows the loading on individual resources. You can see an example of both together in Figure 11-2.

The precedence view, on top, shows a piece of Figure 11-1 for those tasks we are currently dealing with. The resource view, below, is a more standard

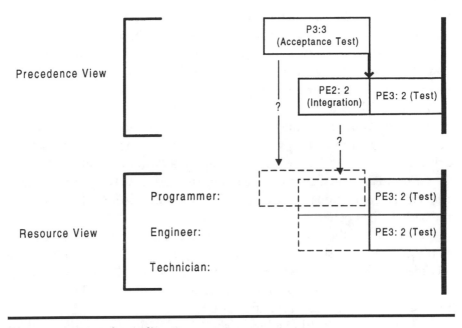

Figure 11-2   Load Leveling Process-1

type of Gantt chart. It has three rows, one for each available resource. The horizontal time scale is the same as for the precedence view, so that the tasks line up vertically. The precedence view is probably a more interesting picture for the project manager; the resource view is probably more interesting to the functional or resource manager.

On the resource view we've had to split out PE3 into two blocks, one for the programmer and one for the engineer, before placing them. The possible placements of PE2 and P3 are shown with dotted lines. The question is, which task should be placed later?

Answering this question is the most difficult part of any load-leveling algorithm. For a single project, you probably want to make the choice that minimizes the overall duration of the project. Unfortunately, you can't really know which choice will be best until you complete the process. Therefore, most computerized approaches iterate through many possibilities to pick the "best."

How much do you care about "best?" It's far better to use a very good approach and implement it well enough, than to use a poor approach and optimize it. Our uncertainty in when we will complete the project is quantified by the project buffer. Therefore, anything that is insignificant next to the project buffer is insignificant. If (for example) a project buffer is

10 months long, then a difference of a month or two in how load is leveled may not be considered significant. Even though we haven't yet determined the size of the project buffer, the guidelines in Chapter 12 will give you some idea of how large it will be. Here we won't worry about it; instead we'll use a simple load leveling procedure that will usually be "good enough."

We would like to make our decision based on which placement will minimize the duration of the final Critical Chain. We don't know enough to do that, but we can make a guess. As an approximation, we guess that the Critical Chain length before the competing tasks will be about the same as the Critical Path lengths for those parts of the network. That means we can calculate an "approximate" Critical Chain duration for each of the choices, and see which is better.

First compute the length of the critical path before (and including) each of the competing tasks; we'll call these CP_1 and CP_2. For P3, there are no precedence dependencies before it, so CP_1 is 3 months. For PE2 we add up the tasks along the critical path up to it, namely E1 → P1 → T1 → PE1 → T3 → PE2. This gives us 13 months for CP_2.

This means that PE2 has a longer chain, and is therefore more likely to make the project late. We place PE2 later, which effectively gives it higher priority. We hope that resource contention for programmers will not cause P3 to be a problem.

Next we can easily place T3, since it's obvious where it goes; there's no resource contention conflict. But now there's another riddle: should P3 be placed next, or PE1?[4] This dilemma is shown in Figure 11-3.

P3 has a path duration before it of 3 months; PE1 has a path duration of 10 months. It looks like PE1 should be placed later. On the other hand, if P3 is placed later, it can be put right next to PE2. If instead we put in PE1, there will be "forced slack" of a month to account for the precedence dependency PE1 → T3. In other words, we seem to have a month of idle programmer time. So if we want to get more sophisticated, we can subtract the "forced slack" from the PE1 chain. It is still much longer, so we still place PE1 later.

Naturally, this process is only a guide. It may be that this 2 months is insignificant relative to the overall project plan, and for some reason you'd rather have PE1 placed earlier. In that case there's no reason why the process shouldn't be overridden.

This process must be continued until all the tasks are placed on the resource view. By scheduling in this way it's possible that we'll have to place a task earlier than the start of the scheduling horizon. That is, by leveling the load in a realistic way we have pushed things too early. That means we must

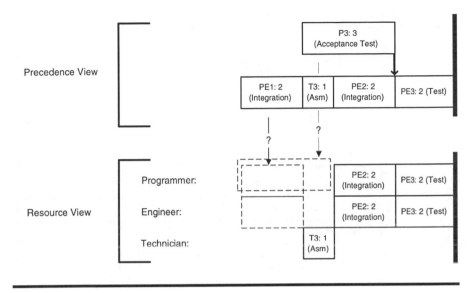

Figure 11-3   Load Leveling Process-2

push all tasks later. (Usually the start of the scheduling horizon is "now." This rule just reflects the fact that it's unrealistic to schedule anything yesterday — despite the fact that people do it all the time.) We shove the entire set of tasks to the right along the time line until the earliest task starts on the scheduling horizon start.[5] At this point we have constructed the picture shown in Figure 11-4.

In Figure 11-4 the load is leveled realistically. The project end date has been pushed out 3 months from January 1998 to April 1998 due to resource contention. Now we're ready to identify the Critical Chain.

Step 3: Determine the Critical Chain

The Critical Chain is that set of tasks which determines the overall duration of the project, taking into account both precedence and resource dependencies. Because of the way we leveled load by pushing to the past, we know that everything is scheduled as late as possible. No tasks can be pushed to the future without pushing the completion date. The Critical Chain tasks are those that also can't be pushed to the past; that is, they have no "give" to either the past or the future. To identify the Critical Chain, we need to compute how far to the past each task could be moved, without pushing anything before the horizon start.

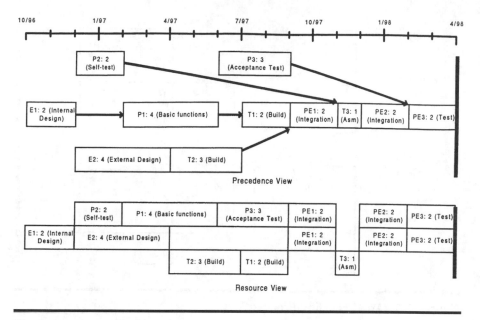

Figure 11-4   Leveled Load

In this picture, that is fairly easy to do. Certainly the leftmost task, E1, is part of the Critical Chain. Examination will show that E1, E2, T2, T1, PE1, T3, PE2, and PE3 form the complete Critical Chain. All other tasks could be moved some amount to the past without going before the start of the scheduling horizon. Marking the Critical Chain tasks with bold outlines, we get Figure 11-5.

Above and to the left of each box on the Precedence View is the aggregate past slack of that task; that is, how far the task could be moved to the past before pushing something before the start of the horizon.

Notice the differences between the Critical Chain in Figure 11-5 and the Critical Path from Figure 11-1. Which picture will be more useful in helping to determine a reasonable project duration?

In the general case, it may not be this easy to identify the Critical Chain. That's because there may be more than one of some resources, in which case it's not obvious how the resource dependencies interact to reduce the aggregate past slack.

If that happens, the simplest approach is to make a copy of your project picture, and move a vertical ruler along the time axis from left to right, starting at the beginning of the scheduling horizon. As a task is encountered, it is moved as far to the left as precedence dependencies and available

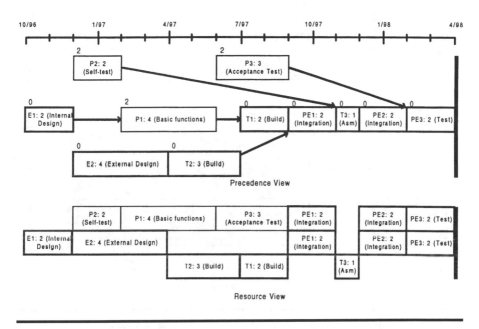

Figure 11-5   The Critical Chain

resources allow. From Figure 11-4, we would get the picture shown in Figure 11-6.

Figure 11-4 gives late starts for tasks; Figure 11-6 gives early starts. By taking the difference between task start times in Figures 11-4 and 11-6, we can come up with the aggregate past slack numbers of Figure 11-5. Those tasks with zero aggregate past slack are the initial Critical Chain.

You might have noticed that the beginning of this section is entitled "Determine the Critical Chain," rather than "Identify the Critical Chain." This is an important distinction. The Critical Chain will determine, to a large degree, what we will focus on. We can't let an algorithm decide by itself, so now we must consider: is this where we want the Critical Chain to be? This is an important strategic decision, for the project and for the company. There are a number of reasons you may choose to change the tasks that make up the Critical Chain. For example,

- There are Critical Chain tasks that can easily be shortened by using more resources. Shorten them, then go back to Step 2, Load Leveling.
- The current schedule puts some high-risk tasks at the end of the schedule. Move them earlier, and go back to Step 2.

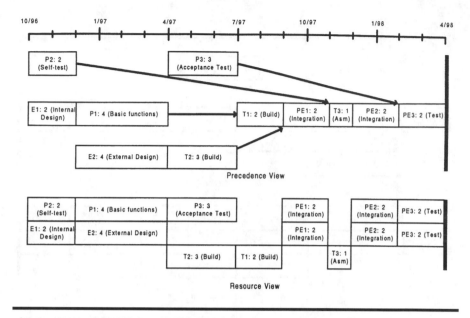

Figure 11-6   Identifying the Critical Chain by Pushing Early

- Some tasks on which you believe the project's completion will depend have not been identified as part of the Critical Chain. Figure out why, and adjust the schedule according to what you believe is right.
- Take into account other considerations. For example, since the Critical Chain tasks will receive the most scrutiny, it may make sense to define the Critical Chain to be tasks over which you have some control.

Any of these changes may require going back to Step 2, Load Leveling, and repeating the scheduling process.

After completing these first three steps, we have a schedule that is realistic, except that it doesn't take into account uncertainty. In Chapter 12 we protect against the uncertainty by putting in the buffers. This will immunize the schedule, thereby making it practical.

Key Concepts

- A Critical Chain schedule is designed to minimize project duration while minimizing WIP.
- The first part of the scheduling process produces a schedule that, other than ignoring uncertainty, is realistic and minimizes WIP.

- The first step in the scheduling process is to create a project plan, with tasks placed at their late start times.
- "Better" decisions are only important if the difference is significant relative to the duration of the project buffer.
- The second step is to level load across resources. The load-leveling process does not need to be "best" or "optimal," just "good enough."
- The third step is to determine the Critical Chain. That is done first by picking those tasks which form the longest chain through the network; and second, if necessary, by changes based on the scheduler's knowledge of the processes.

Questions for Further Thought

1. How could you use the schedule developed to this point?
2. Is it possible to have parallel Critical Chain paths? When would it make sense to allow that?
3. Can you think of types of resources for which it makes no sense to try and resolve resource contention?

12 Adding Buffers

I t's not enough to identify the Critical Chain. Yes, we have a good idea where to focus, but everything else is placed as late as it can be. In its current form the schedule is completely vulnerable to disruptions. Having gotten rid of any extra padding in the task times, we must now put some back to protect our customer. But keep in mind that these buffers aren't really padding, and they aren't a luxury. They are a key factor in on-time completion.

In order to add buffers, we need to go through three steps. First, we must identify the points where the three types of buffers go — the project, feeding, and resource buffers. As we will see, once the Critical Chain is identified that becomes easy. Second, we must decide how large the buffers should be. How much protection do we need? Third, we must reschedule tasks, preferably earlier, to accommodate the buffers; that way the buffers will be available to protect the project.

Step 4: Identify the Buffer Points

Since we're dealing with a single project, we have to identify the locations of three types of buffers: project, feeding, and resource. The project buffer is easy to flag; it goes after the last task(s) in the project. In our case, that would be after PE3. The project buffer is designed to protect the customer, primarily against fluctuations in the Critical Chain tasks. The project buffer may also be used if tasks on a feeding path use up their feeding buffer.

Feeding buffers are placed wherever a non-Critical Chain task joins the Critical Chain, both to protect the Critical Chain from disruptions on the tasks feeding it, and to allow Critical Chain tasks to start early in case things go well. For our example, feeding buffers go between P1 and T1; between P2

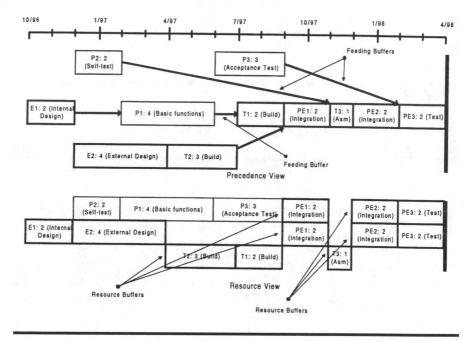

Figure 12-1   Feeding and Resource Buffer Points

and T3; and between P3 and PE3. Feeding buffer points are shown in the precedence view of Figure 12-1. It's pretty obvious that we won't have any problem buffering these last two buffers, given the way tasks have been placed, but we will put in all buffers to illustrate the process.

Resource buffers are placed wherever a resource has a job on the Critical Chain, and the previous Critical Chain task is done by a different resource. The resource buffer points for our example are shown in the resource view of Figure 12-1. The resource buffers are used to make sure the resources are ready to work on the Critical Chain tasks. The easiest way to see where resource buffers belong is by looking at the resource view. Every boldly outlined (Critical Chain) task that is preceded by a non-bold task or empty space needs a resource buffer before it. The technician starting T2 needs a resource buffer before his job. The programmer and engineer working on PE1 need a resource buffer. The technician working on T3 also needs a buffer, as do the programmer and engineer doing PE3. No resource buffer is needed before T1, because this isn't the first consecutive Critical Chain task the technician is working on.

Sometimes there are two different types of buffers apparently in the same place. For example, there are both a feeding buffer and a resource buffer before task T3. These buffers are part of different views, and they have different meanings. The feeding buffer makes sure work is available so that the Critical Chain tasks won't be starved for work. The resource buffer is time during which the given resource should be on the alert to start the next Critical Chain task. Depending on how we decide to treat the resource buffer (see Step 6 below), the resource buffer might just be a wake-up call; we might actually reserve space on the resource during the buffer time. We flag all the buffers in order to make sure we have what's needed, even though, in this example, the feeding buffer before T3 is unlikely to be very interesting.

Step 5: Decide on Buffer Sizes

Statistical fluctuations exist. By definition they can only be estimated. We can't ignore them, and unfortunately we also can't precisely predict their magnitude. Fortunately, even without precision we can put in protective buffers that are good enough to protect the schedule.

The most important buffer is the project buffer, because that is ultimately what protects your commitment dates. The project buffer length should be based primarily on the cumulative risk along the Critical Chain, since it is mainly protecting the customer from Critical Chain fluctuations. The question is, how much buffer time do you think would be needed to give a 90% (or better, if needed) chance of finishing the project on time?

In theory, we can claim that the buffers are some kind of aggregate of the risk encountered along the chain of events feeding them. We could certainly go into great detail with formulas and calculations, worst-case and best-case timings, and so on. This would not be worth the trouble. The data just aren't good enough to support precision or complex calculations.

In practice, we want buffer sizes that are good enough. Traditionally we would have inflated the task time estimates; now we are using averages. We have "saved" a lot of time in the project; now we have to give some of it back. How much? Suppose the original padded project duration would have been 4 years, and it has been pared back to 2. That means the total duration of the Critical Chain is 2 years. Why not start with a project buffer of half of the padding we pruned, or 1 year? We have removed a year from the project duration, and still have a year of protection left. However, this protection protects the entire project, rather than individual tasks in the project. A good rule of thumb is to start with 50% of the (unpadded) Critical Chain duration for the project buffer.[6] Since the Critical Chain duration for the wideband receiver project is 18 months, this initial estimate would give us a 9-month project buffer.

The feeding buffers are needed to make sure that the tasks on which the Critical Chain depends will be ready in time. They also provide the chance for early start, if the Critical Chain tasks are ahead of schedule. Most likely a 90% chance of being ready on time will be sufficient here as well. It's probably valid, again, to set the buffer size at half of the padding saved in the path leading to the feeding buffer.

Assuming we treat resource buffers strictly as wake-up calls, they're easy to size. We could, for example, make all resource buffers 2 weeks long. Then, based on how the Critical Chain schedule goes, we would make sure the resources are notified at appropriate times before they're needed.

An Alternative Calculation[7]

The guidelines given above for sizing buffers are probably sufficient for most purposes. However, there are cases where it pays to be more sophisticated; for example, if risk varies greatly from one task to another, or where the NIH law applies (see Chapter 24).

We can base buffer sizes on the aggregate risk along the chain feeding the buffer. We start by asking for worst-case duration estimates for each task. Let's say we assume a lognormal distribution for the probability of completion of a task. The lognormal distribution has a large hump at the start, and a long tail (see Figure 2-1). Assume also that tasks should be completed within our worst-case duration estimates around 90% of the time. This means the difference between the average expected duration and the worst-case duration will be about two standard deviations. For each task feeding the buffer, if w_i is the worst-case duration and a_i is the average duration, the presumed standard deviation would be $(w_i - a_i)/2$. If we can further assume that the sum of the distributions will be normally distributed (that is, pretend the Central Limit Theorem applies[8]), and assuming we want a buffer that's two standard deviations long, we can say,

$$\text{Buffer} = 2 \times \sigma = 2 \times \sqrt{\left(\left(w_1 - a_1\right)/2\right)^2 + \left(\left(w_2 - a_2\right)/2\right)^2 + \cdots + \left(\left(w_n - a_n\right)/2\right)^2}.$$

This might be the preferred buffer sizing approach for those who want a scientific approach to sizing buffers, such as scientists and engineers.

Step 5b: Adjusting Buffer Sizes

The calculations described above for buffer sizes are very approximate. At this point you may want to adjust your buffer sizes, based on the following considerations:

- How confident are you that your resources will be made available as needed for the entire project? If other late projects will "borrow" your resources or delay their availability, buffers need to be increased.
- How many tasks are in the Critical Chain? If there is one task, the buffer should probably be ($w - a$). In other words, we need to provide for the worst case. The buffer is identical to padding the individual task. If there are many tasks, the sum can be reduced by some factor, because — as in our Saturday morning example of Chapter 8 — some positive and negative fluctuations will cancel out.[9]
- How many complete unknowns are there in the project? Significant risks may not be taken into account sufficiently, and may require bigger buffers.
- What is an acceptable delivery time? What can you live with? If adding the project buffer will cause the completion date to be pushed out unacceptably, a 90% chance of on-time delivery may be too high to hope for. You'll have to accept greater risk; which might be mitigated by more careful monitoring of Critical Chain tasks. On the other hand, if (for example) a multimillion dollar flight test has been scheduled based on your completion date, you had better try for more than a 90% chance of making it.
- Is the chance of early start on the Critical Chain great enough to warrant increasing the feeding buffer sizes?

Ultimately, of course, the decisions on how to size buffers will come down to intuition, experience, and finally, buffer management. Nevertheless, it has to be better to estimate the buffers explicitly and treat statistical fluctuations and dependent events as real, than to buffer all tasks a little and hope the problems don't multiply.

There is a useful rule in making timing decisions for project planning in general, already noted in Chapter 11: **If the decision is about something that is insignificant relative to the size of the project buffer, it doesn't matter what decision you make.**

Our uncertainty for the overall project is at least as much as the project buffer, which means we can use the size of the project buffer to decide what's important. How do we define "insignificant?" Should it be 20%, or 10%, or 5%? Pick one.

For our project, we'll stick with 50% of the feeding tasks for the project and feeding buffers. That gives us 9 months for the project buffer and varying amounts for the feeding buffers.

We will also put in 1-month resource buffers, and we will not push tasks early to accommodate them. In other words, in this example the resource buffers will just be wake-up calls.

Step 6: Insert Buffers

Having identified the buffer points and buffer sizes, finally we must put the buffers into the schedule. The project buffer just moves out the date the customer may expect the project to be completed by. In fact, the end of the project buffer is what we should use for the completion date of the project. The feeding buffer between P1 and T1 requires some tasks be pushed earlier. The resource buffers are shown as short blocks, because they don't really take space on resources. Figure 12-2 shows the complete buffered schedule.

We have put a dotted line on the time line, rather than extending the diagram to the right to show the full project buffer. The buffers are flagged according to their types: F for feeding, R for resource, and P for project. Resource buffers are placed on the resource view because they protect the Critical Chain from resource dependencies. Feeding buffers are shown on the precedence view, because they protect the Critical Chain from precedence dependencies. Note that all buffers except the project buffer are placed flush against the Critical Chain tasks they feed. That's because the resource and feeding buffers protect the Critical Chain, and the project buffer protects the overall project.

We now have a complete, buffered schedule. We've done our best to create a schedule that can be met. We know where to focus attention to make sure the schedule stays on track, namely the Critical Chain and its buffers. We can tell how important each task is in determining the overall completion

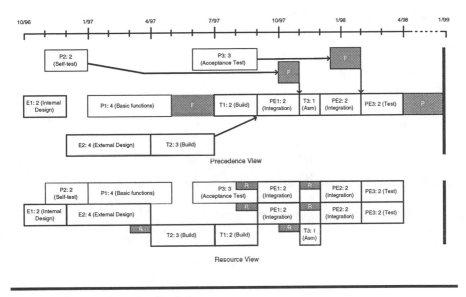

Figure 12-2   Buffered Schedule

date. We can start to consider means of measuring the global impact of changes and other kinds of disruptions on the schedules.

Pushing the Project

Before moving on, there's another question that might come up while inserting buffers: when should the project due date be pushed later to make room for buffers? Suppose first of all that the wideband receiver project is due at the start of the year 2001. That leaves plenty of room to push all its tasks earlier to make room for the project buffer. We would certainly prefer doing that, rather than allowing the buffer to push out the project due date. So pushing earlier is our first choice. If we can't push earlier, it makes sense either to push out the project due date, or decide how we can make the project shorter. The easiest way of making the project shorter seems to be shortening the buffer, but that's the last place to look. The project buffer should only be reduced if there is good reason to assume that a shorter buffer will be sufficient. The project buffer is not slack, it is a key part of the schedule.

What about feeding buffers? Suppose in Figure 12-1 we wanted 4 months of feeding buffer between tasks P1 and T1 rather than 2 months. What should we do? We can push P1 and P2 earlier by 1 month to increase the buffer to 3 months. After that, we seem to have to push out tasks after the buffer. This would push the Critical Chain tasks later, and therefore the project completion date. If the buffer duration is realistic, then this is the only real choice. In this specific case we could argue (according to the rule given above) that the extra month is insignificant relative to the project buffer, and therefore it doesn't matter if we put it in or not. In this case either way works. There will also be times when the project due date must be pushed to accommodate the feeding buffers.

Comparison with Traditional Scheduling

There are a number of ways that Critical Chain scheduling differs from traditional Critical Path analysis:

- "Average" task duractions are used, not "safe" duractions.
- Slack is neither eliminated nor "spread out" through the schedule. Instead, buffers are applied at key points.
- Tasks are placed as late as possible, without risking the project completion.
- Both precedence and resource dependencies are taken into account.

These differences lead to the following major benefits:

- Project lead times are cut significantly, by pooling the "slack" into strategically placed buffers.
- Investment and work-in-process are minimized.
- Project completion dates are secure.
- The need for rescheduling is minimized.
- Task priorities are clear.

Of course, these are just the technical aspects of deriving a schedule. There are many considerations to using such a schedule; implementation issues will be discussed in Part IV.

Key Concepts

- We must identify the points at which to place project, feeding, and resource buffers.

- Buffer sizes can be determined approximately, based either on the average task duration estimates, or a combination of average and worst-case duration estimates.
- Individual buffer sizes can be adjusted based on intuitive assessment of risk.
- Buffer insertion may cause the Critical Chain, and hence the project completion date, to be pushed later.
- The Critical Chain approach to scheduling helps minimize project duration and WIP, delay investment as far as possible, and maximize the chance of on-time completion.

Questions for Further Thought

1. Invent a simple project and try scheduling it using the Critical Chain approach. Sticky papers are useful for sliding tasks around. What questions arise? What principles apply in resolving them?
2. Is it feasible to develop a means of sizing buffers that is completely automated?
3. Inserting feeding buffers may cause gaps on the Critical Chain. What would be the effect of these gaps?
4. What should be the effect on the Critical Chain schedule of having intermediate due dates (milestones) that must be met?

13 Planning the Critical Chain Project

There are still some holes in the planning process, especially Step 1: Create the Initial Plan. In this chapter we'll look at a nonstandard tool, the "Prerequisite Tree," and apply it to creating an initial plan. This tool is frequently used by TOC practitioners as a means of creating implementation plans that achieve specified objectives, and also as a means of overcoming obstacles to reaching those objectives.

We'll also put the Prerequisite Tree and Critical Chain scheduling into a broader context to encompass more of the planning process. We will not attempt to provide a detailed, generic planning methodology; that would take us too far off course. There are wide variations in the levels and formats of planning required from company to company and job function to job function. You might need statements of work, work breakdown structures, milestones, and so on. There are also variations in the management hierarchies involved, including different types and descriptions of higher-level managers, project managers, and functional managers.

Instead, we'll discuss means of addressing some of the shortcomings of traditional project management, with the understanding that the approach presented will need to be adapted depending on the environment. In particular, we will look at aspects of

- Answering "why" as well as "what," "who," and "when"
- Communication
- Creating realistic plans

First, let's start with a definition: **A project plan is a description of a set of activities that, if executed according to the plan, should achieve the objectives of the project.**[10]

Why create a plan? We can list lots of specific reasons; a general guideline might be more useful. We assert that **the purpose of a project plan is to develop and/or communicate understanding of the project.**

Plans are created as a means of assisting decision making both before and during project execution, which means they help to develop understanding. They are created as a means for control, which requires the communication of understanding. Intuition alone is insufficient.

Are there other good reasons for creating a project plan? Some would argue that they create plans in order to fulfill company policies. If those policies do not foster development and communication of understanding, the contents of the plans are probably not useful in achieving the objectives of the projects.

Understanding is key. There are at least two requirements to communicating understanding that we often ignore. First, we must not communicate misunderstanding. If we don't know something, we should not communicate it as truth. If a task might take 5 days and it might take 25 days, it's unlikely that it will take 15 days. If someone says "there's a bomb in your office set to go off between 30 minutes and an hour from now," you won't act on the assumption that you have 45 minutes to get out. We should base our actions on the knowledge and precision we have available.

Uncertainty is a dominant factor in planning and managing projects. There is uncertainty in how long tasks will take, when and how many resources will be available to accomplish them, and often even uncertainty in what needs to be accomplished. Some managers feel that, since plans can never be "precise," there is little point in making them. When plans are made they are frequently treated as unimportant or even irrelevant. Some others, on the other hand, feel that plans should always be as precise as possible, even if that requires making guesses about what's uncertain, and pretending the guesses are exactly correct.

Of course, "precise" does not mean "correct," and the opposite of "precise" is not "wrong." There is always knowledge we have and knowledge we don't have. It is extremely important to organize and communicate both kinds of knowledge. What we know provides the only rational basis we can have for making decisions. What we don't know defines what we'll have to find out by the end of the project. The more uncertainty we start with, the more important it is to make a plan and gather the knowledge we have; so that we do the best we can, without mistakenly assuming knowledge we don't have, or making decisions we don't have the ability to follow through on. We need buffers, and hence Critical Chain scheduling, to help quantify the imprecision, to separate the known from the unknown.

In addition, "what" is usually insufficient to communicate our understanding; a second requirement to communicating understanding is explaining "why." Suppose, for example, you want to take a picture of two 3-year-olds. In order to get a cute picture, you say "put your arms around each other." Most likely each child will obediently fold his arms across his chest. They have carried out the instruction perfectly. And yet if the children understood the reason for the request, their actions would be very different.

Also remember that "why" can have local and global meanings. The immediate reason for a task is important; the overall (global) objective it is intended to meet is also important. Lack of understanding of either can lead to wasted time and effort. The Prerequisite Tree is useful in communicating the "whys" associated with a project plan.

We propose a planning process that contains the following kinds of elements:

1. Clearly state the objectives of the project and of the project plan.
2. Determine the needs to be met and the tasks needed to meet them.
3. Determine the logical relationships between tasks and needs.
4. Estimate the resource requirements, task durations, and costs.
5. Calculate the Critical Chain schedule, including buffers.
6. Evaluate the plan according to budget and timing restrictions.
7. If necessary, go back to an earlier step and revise the plan.

Let's go over these steps in detail and look at what makes them useful. We'll look at the steps in the context of a simple project. Jack is planning on building a treehouse for Kerry and wants to plan his project.

Step 1: Clearly State the Objectives of the Project and of the Project Plan

This is a very important step, yet we often assume at least part of it. Frequently the project objectives are thought of in terms of the customer specifications. For example, the objective for a project to develop a radar warning system might be specified as "Meet the requirements laid out in document XYZ-55089." That says the objective of the project is every little piece of document XYZ-55089. In reality, the objective (as defined by the customer) is probably something more understandable, such as "provide three radar systems that work in the field by a particular date." Don't confuse a specifications document or a statement of work with the project objectives needed to satisfy the customer.

The reasons for constructing the plan need to be clearly verbalized as well. Let's consider some of the reasons people need plans for projects. A project plan can be useful to:

- Evaluate whether the project makes sense to undertake.
- Evaluate the impact of this project on other projects.
- Enable people to coordinate their activities.
- Sell a project internally.
- Sell a project externally.
- Prioritize resource allocation among different projects.
- Assign jobs to specific workers.
- Monitor project status.

Notice that all these reasons have to do with developing and/or communicating understanding. It follows that when you state the objectives of a project plan you must describe whose understanding you're talking about, and what kind of understanding they need from your plan.

The objective of Jack's project is to build an acceptable treehouse. Does this seem clear? If we examine it carefully, we notice that the word "acceptable" is definitely open to interpretation. It must be acceptable to Kerry, or she won't play in it. In terms of safety, it must be acceptable to Janet, or else she won't let Kerry play in it. It must be acceptable to Jack in terms of time and effort, or he'll never finish it. Let's clarify the project objective by saying that it is to "build a treehouse which meets Jack's, Kerry's, and Janet's requirements." Any remaining vagueness will have to be resolved as part of the project plan.

The objective of Jack's project plan is first, to develop an understanding of what needs to be done. He wants to structure his actions so that he doesn't leave anything out; and so that he gets a sense of how long the project will take. He would also like to use this plan to communicate this information with Kerry, so that she can know when to expect a treehouse, and with Janet, so that she knows how much time Jack will be unavailable for raking leaves or fixing faucets.

Step 2: Determine the Needs to be Met and the Tasks Needed to Meet Them

Here we make a distinction between "needs" and "requirements." A requirements document might be referred to in the project objective. By "needs" we mean specific items that we know must be accomplished. A need could be an obstacle that must be overcome, or it might be a requirement that will map to a specific task.

The purpose of your plan, whether responding to a request-for-proposal, making a long-term plan, or setting up short-term schedules, will have a big impact on the level of detail of scheduling. It seldom makes sense to go into greater detail than "good enough." In fact, it's easy to mistake greater detail for better information, which can lead to errors in judgment. A car priced at $12,985 gives a buyer the impression that the dealer knows just what they want; there is precision and no uncertainty. A price of $13,000 implies that there's room for negotiation. Relative to the price of the car, the difference is negligible. But a consumer can make a big mistake by assuming that one number is somehow more precise.

You must decide how precise to be in laying out tasks. Better than "good enough" may be misleading detail. "Good enough" will depend on the objectives of the plan, and the type of project. The needed level of precision must be determined in a few areas:

- Aggregation of resources: "two programmers" is less precise than "Bill and Ted." Ten people for 4 years is different than 40 people for 1 year.
- Detail for timings: times can be laid out hourly, daily, weekly, or monthly. For most projects hourly is far too detailed.
- Clarity: tasks can be explained in different levels of detail. Probably you will start at a high level, and refine the individual pieces later as needed.
- How many tasks can you manage? Usually more than 100 becomes extremely unwieldy, and you will need to do multiple levels of planning.
- Transfer batches: In so cases smaller tasks help clarify that work can be done in parallel.

Frequently tasks are lumped together during high-level planning when calculating project duration. This may be the best thing to do. However, it can also inflate the apparent time needed for the project. All too often aggregated tasks must be split apart for detailed scheduling, but the possibility of overlapping the pieces is not taken into account. This can inflate the time actually required for the project.

Table 13-1   Tasks, Needs, Obstacles, Objectives for Building a Treehouse

Build a treehouse that meets Jack's, Kerry's, and Janet's requirements. (Objective)
Make sure the treehouse is safe. (Need)
Cut the lumber. (Task)
Jack doesn't have lumber or nails. (Obstacle)
Design the treehouse. (Task)
Assemble the treehouse. (Task)
Measure space available in the tree. (Task)

For example, the design phase of a project might require design of two major pieces of hardware in the system. If design is considered to be one task, the possibility of designing one piece, then building it while designing the other won't be considered. This was also shown in Figure 3-3. For planning purposes, depending on the level at which you're planning, you may need to group tasks together. For detailed schedules, **split tasks when significant time can be saved by overlapping them in time and/or across resources, or when splitting is required for clarity. Otherwise keep them together.** In other words, we should gain significantly by splitting, either in clarity or through shortening the time to complete the project.

Each task that we "feel" is necessary has reasons that it's needed, or needs that it fulfills. Each need must be fulfilled by one or more tasks. Start by listing all the tasks and needs, without necessarily associating them. We will later associate tasks and needs to form a table, and eventually show each task meeting one or more needs, and each need fulfilled by one or more tasks.

The final table will tell us not only what is needed, but why. This exercise is very useful in understanding and communicating how tasks fit together to make the overall project work. It is also important for the next step, determining the logical relationships between tasks. If we don't know why we're carrying out tasks, the relationships we assume between them may not be logical.

For Jack's project, we can start out with the list shown in Table 13-1. Then we convert the list in Table 13-1 to tasks and their associated needs/obstacles, filling in any missing information (Table 13-2).

As Jack looks at the list in Table 13-2, certain tasks might seem difficult to start with. For example, the design might be a problem, because Jack is not an architect. Right now he's not considering hiring someone, as that would no doubt be very expensive. He's planning on finding a book describing

Table 13-2 Objective: Build a Treehouse that Meets Jack's, Kerry's, and Janet's Requirements

Task	Need/Obstacle
Research safety requirements.	Make sure the treehouse is safe.
Cut the lumber.	Standard-sized lumber doesn't have the required dimensions.
Buy materials.	Jack doesn't have lumber or nails.
Design the treehouse.	There must be specs to build to.
Assemble the treehouse.	A final treehouse must be constructed.
Measure space available in the tree.	The treehouse must fit in the tree.

how to build treehouses. This suggests another task: buy a book about building a treehouse. Perhaps this is too much detail for the purpose of the plan, but it certainly helps clarify what he's doing. The need is "Jack needs to learn about treehouses." This also ties in with the safety issue; learning about treehouses will probably help determine if the treehouse is safe. Notice that the task only needs to be sufficient to meet the associated requirements. It doesn't have to do anything more. Jack doesn't need an architect to write treehouse specifications.

We also might notice something else missing: Janet and Kerry haven't been involved. Jack doesn't know their requirements. This suggests another task: meet with Kerry and Janet to determine their requirements. The need this fulfills can also be expressed as an obstacle: Jack doesn't know what Kerry and Janet want.

Some of the needs seem unnecessary. "Cutting the lumber" must be a task, but the need is pretty obvious. Needs can be omitted for certain tasks, but that's usually not necessary. There's not much payback in being lazy with easy things. This need is easy to verbalize, and it even exposes an assumption: Jack could buy a prefabricated treehouse, and the step would be unnecessary.

Step 3: Determine the Logical Relationships between Tasks and Needs

This is a two-part step. First, order the tasks in time sequence, to get the approximate flow of the project. This doesn't need to be precise; we'll make it more rigorous in the second part. We'll put the last tasks at the top, because that way they'll be nearest to the final objective (Table 13-3).

Table 13-3 Time Sequence for Building a Treehouse

Task	Need/Obstacle
Assemble the treehouse.	A final treehouse must be constructed.
Cut the lumber.	Standard-sized lumber doesn't have the required dimensions.
Buy materials.	Jack doesn't have lumber or nails.
Design the treehouse.	There must be specs to build to.
Measure space available in the tree.	The treehouse must fit in the tree.
Research safety requirements.	Make sure the treehouse is safe.
Meet with Janet and Kerry.	Jack doesn't know what Kerry and Janet want.
Jack buys a book about treehouses.	Jack needs to learn about treehouses.

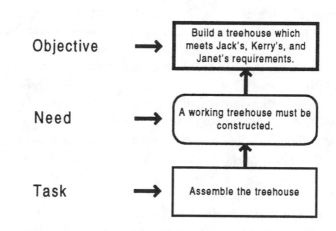

Figure 13-1 Tasks/Needs Connections

The second part of this step is to arrange the tasks and needs (or obstacles) into a logical structure. This structure is sometimes called a "Prerequisite Tree"[11] because it links all the prerequisites needed to achieve the objective. We must connect the tasks and requirements together in a way that expresses precedence requirements between tasks. The structure also tells why each task is needed. We start with the objective, look for the last task that seems to be required, and connect up from the task to the requirement and then to the objective. For example, see Figure 13-1.

Notice in Figure 13-1 that the project objective is in a bold box, to set it apart. The needs or obstacles are in rounded boxes, and the tasks are in rectangular boxes. The diagram is read from the top down, "in order to ⟨top square box⟩ I must ⟨bottom square box⟩ because ⟨middle round box⟩." In this case, "In order to build a treehouse which meets Jack's, Kerry's, and Janet's requirements, I must assemble the treehouse, because a final treehouse must be constructed." This example might seem a bit silly, because treehouses don't assemble themselves. On the other hand, it might also help you to notice that something fairly important is missing from the bottom box: the word "install."

The way you read the statements doesn't have to be precise, as long as it helps to examine the logic. For example, "I must" could change to "I must have," and "because" could change to "to overcome." You can also try "I can't have ⟨top box⟩ without ⟨bottom box⟩ because of ⟨middle box⟩." You would then have to change the statements a little. For example, "I can't have a treehouse which meets Jack's, Kerry's, and Janet's requirements without assembling and installing the treehouse, because a working treehouse must be constructed." Let's look at the full Prerequisite Tree based on Table 13-3 (see Figure 13-2).

Read through the tree from top to bottom and see if the links make sense. Reading might result in further rewording. For example, we might decide that "Jack buys a book about treehouses" doesn't meet the requirement "Jack needs to learn about treehouses"; perhaps "Jack buys and reads a book about treehouses" is better.

All tasks and needs for the project must link together to meet the project objectives. If there are needs or tasks that don't fit in this structure, either you need to revise the project objectives, or eliminate the task or needs.

Step 4: Estimate the Resource Requirements, Task Durations, and Costs

Add time estimates and resource requirements to the list of tasks from Step 2, using available experience and historical data if appropriate. The times specified should be expected average times, rather than padded average times. This means that experience and history may be poor indicators for timings. Using expected times is often a difficult cultural switch, because most organizations are used to task durations being padded. It's critical that everyone involved with the plan understand that the times that happen in reality can

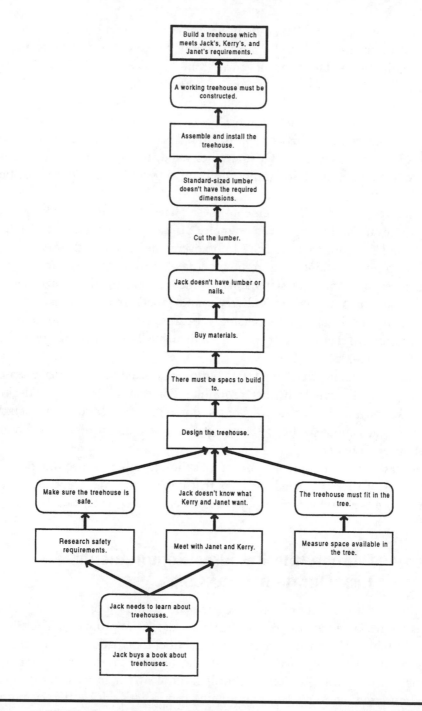

Figure 13-2   Treehouse Prerequisite Tree

be both less and more than the average. We don't want padding. We don't want minimum times either, for two reasons. First, we'll want to use Critical Chain schedules to look at resource requirements, and underestimating durations will underestimate overall resource requirements. Second, if task durations are on average understated, buffers — especially project buffers — will have to be increased to account for this. Given the buildup of errors in scheduling to short durations, the farther you go to the future, the more buffers will have to be overestimated. Any semblance of realism in the schedules can be lost.

Very often someone looking at a schedule can interpret time estimates as either conservative or liberal, depending on their point of view. That happens when people aren't careful to distinguish between elapsed and dedicated time. Usually, time estimates will appear conservative if they're interpreted as elapsed and liberal if interpreted as dedicated. For example, you might expect a task that would require 1 day of dedicated work to take a week when other task commitments are factored in. The time estimates we're looking for here should be *dedicated*. If we take the point of view that the resource is dedicated to the tasks, the time estimates should not appear either liberal or conservative. We are assuming **no multitasking**. The conversion to an elapsed time should be a function of the scheduling process, in particular the resource load leveling step. That conversion may have to take into account other, ongoing duties such as meetings and vacations.

Task durations can be added to the table of tasks and requirements; for example, see Table 13-4. Jack has specified expected or "average" durations; in other words the times he thinks the tasks should take. He hasn't bothered to specify worst-case durations, because he doesn't know enough to predict them.

Table 13-4   Duration, Resources, and Costs of Building a Treehouse

Task	Average Duration	Resources	Cost
Assemble the treehouse	8 hours	Jack	$ 0
Cut the lumber	4 hours	Jack	0
Buy materials	3 hours	Jack	100
Design the treehouse	12 hours	Jack	0
Measure space available in the tree	1 hour	Jack	0
Research safety requirements	2 hours	Jack	0
Meet with Janet and Kerry	1 hour	Jack, Janet, Kerry	0
Buy and read a book about treehouses	3 hours	Jack	30

The "costs" are strictly the truly variable costs for the task, in this case the materials and book costs. If Jack decided to hire someone to help, those costs would be included, since they're also variable. It could be that Jack needs more precision in his cost estimates. For example, he may need to make some calls or a trip to the hardware store to be sure he can afford this project.

Having completed this step we have an initial plan, which is the basis for creating a schedule.

Step 5: Calculate the Critical Chain Schedule, Including Buffers

This process was described in Chapters 11 and 12. It involves creating schedules and adding buffers, taking into account the finite capacity of the resources involved. We'll need to take into account intermediate due dates, if any. Since resources, especially key ones, are often shared between projects, most likely we should take into account the loading across all the interconnected projects. That is, Critical Chain analysis must be done to incorporate the requirements of the new project into existing requirements and capacity. Otherwise, existing projects and the new one may jeopardize each other. We will discuss this more in Chapters 19 and 20.

Our example is simple enough that there's no need to write down the Critical Chain explicitly. Jack is performing all the tasks, so they are all on the Critical Chain. Nevertheless, the different tasks depend on one another, and there will be statistical fluctuations. Even though some early tasks may balance some late ones, there is need for a project buffer.[12] We can use the first approach suggested in Chapter 12 to derive a project buffer size. If we add up the average durations, and then divide by two, we'll get a good enough estimate. The total of all the durations is 34 hours. We're looking at 34/2 = 17 hours for a project buffer size. Since Jack is doing the work himself, the overall estimated completion time will be 34 + 17 = 51 hours.

If Jack works 16 hours each weekend, the project should take about three weekends. The whole family will have to check their other weekend commitments to see if this is possible, since Kerry and Janet may need Jack's time. Conflicts may mean that the project will take longer.

Step 6: Evaluate the Plan According to Budget and Timing Restrictions

Is the project going to finish too late? In some cases projects can be speeded up by purchasing more resources. In many cases they can't.[13] In Part IV we'll look at some other ways of speeding up projects.

Is the cost too high? A tradeoff may be required between cost, project duration, and functionality. Alternatives may need to be explored.

Sometimes there are restrictions that you're not interested in dealing with but must, such as overhead allocations or a company policy that prohibits payment of overtime. Many times you will have to take unpleasant restrictions into account. Be careful; sometimes we assume that these restrictions are sacred, even when the powers that put them in place are amenable to reason.

In our example, once Jack factors in his personal calendar he may decide that the project will take too long. He may recruit friends or hire someone to help. He may also have to scale back on what Kerry wants in a treehouse, perhaps forgoing the chintz draperies and ornamental woodwork. He may also decide to invest more time in finding out the minimum requirements so he can decide what is "good enough," and save the enhancements for another day and another project.

Step 7: If Necessary, Go Back to an Earlier Step and Revise the Plan

If the plan resulting from Step 6 is good enough, this step isn't necessary. On the other hand, if Jack decided to hire someone to help, he might go all the way back to Step 2 and replan the entire project. If he decided just to cut back on some materials, he'd probably only need to go back to Step 4. Repeat steps as often as necessary to produce a good plan. It's much easier to redesign a project before it's built than after.

Complex Projects

Complex projects can have many levels of tasks and subtasks. Often large projects such as building a ship or a telecommunications system require hundreds or even thousands of tasks and people. Typically these are planned and organized in levels, with more and more detail at lower and lower levels.

When using Critical Chain scheduling, the key rule is "keep it simple."[14] That is the only way to stay focused on the big picture. To set up and maintain a project schedule with thousands of tasks is unrealistic. Instead, plan the entire project at a level that has a manageable number of tasks; probably under two hundred. Resource use can usually be planned at this global level. Individual tasks in such a global plan can then be planned with their own Critical Chains. It's possible that lack of information means they can't be planned in detail until close to their start time. It's unlikely that a detailed plan will be needed until then anyway.

Key Concepts

- Project plans help develop and communicate understanding among and between workers and management.
- Develop objectives for projects and for plans.
- Match tasks with the needs they fulfill or the obstacles they overcome.
- Develop a Critical Chain schedule.
- Evaluate the results and decide whether to make changes.
- Keep it simple.

Questions for Further Thought

1. Suppose you work for a company that manufactures widgets, and you are in charge of planning the building of a new production facility. What would be a reasonable objective? What would be some major needs? Work through Steps 1 to 3 above.
2. Pick a task that you are currently working on. Write the needs, obstacles, and tasks, and put them into a Prerequisite Tree.
3. Suppose you have been given a project to manage that has too many tasks to deal with comfortably. What should you do?

Endnotes

1. Parkinson Jr.'s Law: Play expands to fill the time available.
2. For more information about the "Evaporating Cloud," see references given for the Current Reality Tree; also Goldratt, E. M. and Fox, R. E., *Laying the Foundation, The Theory of Constraints Journal*, Vol. 1, No. 2, 1-6, 1987; and Goldratt, E. M., *What Is This Thing Called Theory of Constraints and How Should It Be Implemented?*, North River Press, Croton-on-Hudson, 1990, Chap. 4.
3. Typically "slack" and "float" are used to describe time that tasks can be made later without affecting subsequent tasks ("free float") or the overall project ("total float"). Since here tasks are placed at their late starts, we'll use the additional terms "past slack" to mean time that a task can be made earlier without affecting the preceding task, and "aggregate past slack" to mean time that a task can be pushed earlier without it, or one of its predecessors, being pushed before the start of the scheduling horizon.
4. In a computerized system, we might need to decide between many tasks on a given resource when figuring out what to place next. In this situation, we can see that the decision is between P3 and PE1.
5. Of course, tasks that do not depend at all on the early tasks may not need to be pushed later.
6. See Goldratt, E. M., *Critical Chain*, North River Press, Great Barrington, IL, 1997, 158–160.
7. Thanks to Tony Rizzo of Lucent Technologies for this approach.
8. The Central Limit Theorem asserts, among other things, that the sum of a large number of independent random variables with identical distributions will be normally distributed. Since typically none of these assumptions (large number, independent, identical) is valid, this analysis becomes just a "reasonable" way to get a buffer size.
9. This points out two problems with the standard probabilistic approach to PERT/CPM scheduling: the buffers are per task, rather than for the entire chain, and the standard formulas for completion probability and risk assessment assume that the probability of completion has a particular distribution. Of course, our complicated buffer calculation has the same kind of problem.

10. This is somewhat different than the standard definition: "A formal, approved document used to guide both project execution and project control. The primary uses of the project plan are to document planning assumptions and decisions, to facilitate communication among stakeholders, and to document approved scope, cost, and schedule baselines. A project plan may be summary or detailed." Project Management Institute Standards Committee, *A Guide to the Project Management Body of Knowledge*, Project Management Institute, Upper Darby, PA, 1996, 168.

11. Dettmer, H. W., *Goldratt's Theory of Constraints*, ASQC Quality Press, Milwaukee, WI, 1997; Goldratt, E. M., *It's Not Luck*, North River Press, Great Barrington, MA, 1994.

12. Note: the most important buffer is the project buffer. Even if it's impossible to identify a Critical Chain, always put in a project buffer.

13. This was described in Chapter 4, in the section called "Using More People."

14. The well-known acronym is KISS, or "keep it simple, stupid."

GLOBAL VIEWPOINT, GLOBAL LEVERAGE

The Critical Chain approach presented in Part II is a powerful technique for planning single projects. But in order to achieve all the miracles from Chapter 7, we need much more. Next we look at Miracle 2: People are focused on global (system-level) improvements rather than local ones. The question, in a nutshell, is how an organization as a whole can become and remain competitive; how it can keep up, or stay ahead, or be wildly successful. How can we engineer significant improvements, where significant is a better bottom line by tens of percent rather than tenths? As we shall see, this question can't be answered by pointing to small, detailed fixes everywhere; instead it requires a global viewpoint, fixes to the entire organization.

The global viewpoint is not just important for those organizations that have an immediate need to improve. No one can afford to be complacent; change is too rapid in today's world. The unsuccessful or marginally successful company usually recognizes a need to improve, although there is frequently resistance to change, as people fear to venture into the unknown. Not surprisingly, the successful companies tend to be the least interested in change. Success breeds complacency, an attitude of "if it ain't broke don't fix it." Any fear of the unknown then magnifies this complacency, and gives it an ability to endure well into hard times. There seems to be a conflict between promoting change in order to be successful long-term vs. promoting stability now in order to avoid short-term instability.

This conflict prompts us to look for changes within the system, rather than changes to the system. We must have layoffs, or reduce the need for rework, or guard the supply room, or have more detailed specifications or measurements. These are changes within the system, and in most cases will not result in significant system-wide improvement. If these changes have a major impact on people within the company, and they produce only small results, it seems as though a major change to the system must be wildly destabilizing.

The conflict must be resolved in two parts. First, we must realize that significant changes are, first of all, changes inside our heads, and don't necessarily imply actions. The real need is for changes to our understanding. If our understanding is broadened sufficiently, the needed changes will become inevitable.

Second, a large part of the fear of change is that we somehow expect radical change to require radical actions. The Gordian knot must be whacked in half with a sword; the company must be re-engineered. Part of the needed understanding is that we must look for areas where small changes will have radical consequences. These areas can be considered an organization's constraints, those areas that limit its ability to improve. They can also be considered its leverage points, because they are the areas where the improvement efforts are the most highly leveraged. If we have to take actions that change everything at once, we will almost certainly cause more problems than we solve.

At this point project managers might wonder how this relates to their jobs. It is traditionally the job of higher-level management to worry about global improvements. In reality, every organization is an interconnected chain. The better the links of the chain can support one another to achieve the organization's goal, the better the organization will perform. The better all the people in an organization understand this concept, the better they will be able to support one another. This part is designed to help people understand better how to help the organization work as a chain, in order that the organization as a whole can improve. This focus of individuals on the entire chain, rather than on their own links, is very important. To become effective, it usually requires a big change in individual mindsets and corporate culture.

In this part we cover the following topics:

- The meaning of the global (system-level) viewpoint.
- The meaning of "improvement" relative to the system.
- Examples of leverage points.
- A powerful five-step process for managing leverage points.

14 | The Goal

The System-Level Approach

The Critical Chain approach to project management is an "improvement" methodology. When I'm sitting at my desk, improvement might mean helping finish a project more quickly, or finding a better-paying job; when I'm being chased by a dog it means something entirely different. For most project managers, an "improvement" might be something that allows more projects to completed, without adding significant expense, or that allows projects to be completed on time, within budget, and accomplishing what the customer wants. This is a fairly specific definition. It may be sufficient for what we've discussed so far, but it is not generic. It is not sufficient when we need to talk about improvement to an entire system.

In order really to effect change, we must look for more than just some rules and techniques. We need to understand more fundamental underlying principles, so that we can respond appropriately to new situations. To start with, we must understand more rigorously what we're improving and what we mean by improvement.

We want to take a **system-level** approach to improvement. That means the "thing" to be improved is a system, and our attention will be directed toward improving the system as a whole. Unfortunately, there are even more definitions of "system" than there are of "project." Here we'll define a system to be anything that is useful to look in the way shown in Figure 14-1.

The system is represented in Figure 14-1 by the box. It has various inputs, various outputs, and probably some intrinsic value, which will depend on one's point of view. For example, a family can be considered a system. As input it may require time and attention; as output it can produce, among other things, love, security, and a sense of self-worth; and it has some intrinsic value, not just in the people but in the way they're changed and, hopefully, enhanced by the interaction. An automobile can be considered a system. We

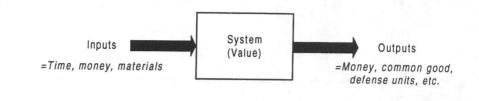

Figure 14-1   Generic System

input maintenance, gas, oil, more maintenance, loan payments, etc.; as output we (the owner) might want reliable transportation or social status.

Practically speaking, you can't talk about a system without having a point of view. How you value your family will differ from how anyone else values it. Fortunately, most people involved with a given system will have some overlap in their views of it. When talking about organizations we'll take the point of view of the owners, because the owners usually have the most say in determining what the organization will do, how it will be run, and whether it continues to exist. In order to understand why a company exists, you must ask the owners. We'll start by looking at the for-profit organization. (For purposes of this discussion, "for-profit" is not a legal entity but a statement of purpose. In terms of legal designations, there are organizations designated "non-profit" that exist to make money, and organizations designated "for-profit" that exist for other purposes.)

Why are for-profit organizations created? By our definition, a for-profit organization exists to make money. Its main purpose or **goal** can be described as making money now and in the future.[1] If it's a publicly owned company, this goal is clear: the stockholders invest in order to make a profit. From the owners' point of view, the company exists to make money. Typically the stakeholders associated with the organization — those involved with it in some way, such as customers, employees, vendors, the government, as well as the owners — will impose necessary conditions. Customers demand quality; employees may demand a certain level of wages, working conditions, or employment guarantees; the government may impose regulations. These necessary conditions can become of overriding importance if they are close to being violated. But the purpose of the for-profit organization, its reason for existence, is to make money now and in the future.

Let's look at the for-profit organization from that point of view. In Figure 14-2 the inputs are broken out into a few categories. Under "materials" we include things you must buy in order to create products, and that can be

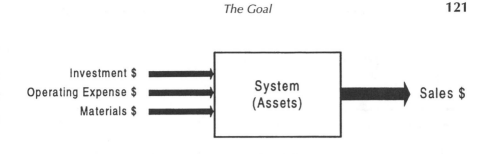

Figure 14-2   Money-Making System

assigned directly to specific products; this category is equivalent to the "truly variable" costs. They are costs that truly vary directly with the quantity of product produced.

Operating expense represents the rate at which money must be put into the organization to keep it running, including salaries, electricity, and equipment maintenance. It is a rate, because any number represents expenses over some span of time. **Investment** is money that is trapped inside the system, again over some period of time. Inside the box, part of the value of the organization, are assets. On the balance sheet, assets are related to investments through depreciation or appreciation. Depreciation is normally reclassified as operating expense for accounting purposes.

The outputs in Figure 14-2 are mostly sales dollars. We define **throughput** to be the rate at which the organization generates money (through sales). In other words, from the above picture, it is sales minus materials, over some period of time. "Materials" actually includes all truly variable costs; that is, costs such as sales commissions, which really vary directly with product sold. We can also include asset appreciation (such as interest income) in throughput, although it's usually not a significant part.[2] For a project manager in a construction company, the throughput associated with a construction project might be the selling price of the project, minus costs of building materials and overtime, minus any penalties (and plus any bonuses). It would not include worker salaries, which are operating expenses, unless the workers are hired through outside contractors. It would not include the new backhoe, which is an investment.

For the sake of simplicity, cash isn't shown as part of this system. It generally comes from clients into the system as a result of delivery of product, then flows out to pay expenses, dividends, etc. The arrows into the box represent things that are purchased; they are a result of paying cash. The arrows going out represent shipped product; they result in cash from customers.

Figure 14-3   Basic Measurements

There's a direct relationship between the values of throughput, operating expense, and investment with the traditional measures of net profit and return-on-investment. Quite simply:

Net Profit = Throughput – Operating Expense

and

Return-on-Investment = (Throughput – Operating Expense)/Investment.

Given these traditional definitions of the bottom line, we're finally in a position to define an improvement for the money-making system. An improvement is anything that increases net profit and return-on-investment. If one measurement goes up, and the other goes down at the same time, we must evaluate more carefully. This may or may not be an improvement.

We can say an improvement will increase throughput, decrease required investment, and/or decrease operating expenses.[3] We want the numbers to move in the directions shown in Figure 14-3.

Not-for-Profit Organizations

Many organizations have a goal that is not to make money; that is, they are not for profit. In such cases, throughput isn't defined in terms of money. Throughput must be measured according to the purpose of the organization. For the military, one might speak of some kind of "defense units" or "readiness." In the medical world, the best-known throughput measurement is quality-adjusted life years (QALYs).[4] The QALY is a numeric measurement of the utility of medical services. It is the expected quality of life given the service, measured on a 0 (death) to 1 (fully normal life) scale, times the expected number of years of life, minus the quality times years expected without the service.

In order to find the purpose of any organization, one must ask the owners. Assuming necessary conditions are met, the owners determine what the company does, and hence what definition makes sense for throughput. A health care organization might exist to make money; it might exist to save lives. A military organization might exist as a defensive force; it might exist to take over the world.

It's not always necessary to quantify throughput. Often the act of taking a system-wide viewpoint, defining the goal and thinking in terms of throughput, investment, and operating expense, can by itself be sufficient to point the way toward improvement. We'll see examples later. Looking back at the traditional measures, net profit doesn't make much sense in a not-for-profit organization, because throughput and operating expense are expressed in different units. However, if throughput is quantified it is possible to look at productivity:[5]

$$\text{Productivity} = \text{Throughput/Operating Expense}$$

Understanding the Goal

Consider the following statements:

- The university would be a much nicer place to work if there were no students.
- The military would do a much better job if civilians would stay out of the way and let them do their jobs.
- My health care plan is great, as long as I don't get sick.
- Why does management keep bothering us? They don't know anything about software development.

There is a common thread here: the viewpoint of the writer seems to be different from the viewpoint of the owners of the system. There is in each case an implied mismatch of goals between those working inside the system, and those who use or run the system. We are tempted to ask, "What are these organizations for, anyway?" What is the goal of a university? What is the goal of the military? If you don't know what the goal of an organization is, it is not possible for you to tell if it is achieving its goal, and it's unlikely that you will work effectively toward that goal.

If you wish members of an organization to contribute toward the goal, they need to know the goal and be encouraged to work toward it. If, for example, the goal of a given university is to provide as much education as possible for its community, and all the employees understand and agree to this, the first statement above should be impossible.

Understanding of the goal is a powerful and essential means of focusing efforts. It is equivalent to keeping your eye on the ball in baseball, or your attention in the center of the board in chess.

Key Concepts

- Throughput, operating expense, and investment are fundamental bottom-line measurements.
- Throughput = Sales – Materials.
- Net Profit = Throughput – Operating Expense.
- Return on Investment = (Throughput – Operating Expense)/Investment.
- The "goal" of an organization is important to verbalize. It is a basic means of focusing the organization.
- The goal of a for-profit organization is to make money, now and in the future.

Questions for Further Thought

1. Try to think of the goals of various organizations you have worked for. Was the goal clear? How did the clarity (or lack of clarity) of the goal affect your job?
2. Think of "awareness of the goal" as a leverage point. How could this leverage point be used to produce dramatic improvements?
3. What is the value of an "improvement" that does not relate to the organization's goal or necessary conditions?

15 Throughput: *Ichiban*

W e now go into some more depth regarding the fundamental measurements, especially throughput and operating expense. If these measurements are really fundamental, they must be thoroughly understood in order to comprehend the meaning of improvement. That, in turn, is a prerequisite to understanding the implications of any improvement process. Therefore, if people are to work together as a chain and contribute meaningfully to improvements that affect the goal of the organization, they must understand the fundamental measurements.

We can hope that these measurements are not equal in significance. We'd prefer not to spread all our energies toward increasing throughput and decreasing investment and decreasing operating expense, all at the same time. This would mean we'd have to consider everything, all the time. Anything that diffuses our attention in such a way cannot be considered a leverage point. What should be the priorities of these measurements? That is, where should we focus our improvement efforts?[6]

The answer, concealed in the title of this chapter, is that throughput is *"ichiban,"* number one. That answer is very important because it means the difference between looking for more work to perform and sell, the throughput approach, vs. looking for ways to spend less, the cost approach. As we shall see, this difference has major ramifications for everyone in the organization.

The Traditional Priority of Measurements

First let's consider what the priorities have been traditionally. What is the most common reaction of management to excess capacity? Cut costs. In times of market downturn, people are usually laid off. Everyone contributes to costs; costs can be attacked by everyone in the company. Reductions can be achieved quickly. The entire system of product costing and pricing is built around this; in fact, a fundamental assumption of cost accounting is that many operating expenses can be treated as variable costs. Traditionally, operating expense is the first place management looks to improve the bottom line, and is therefore number one. Throughput, with its direct impact on net profit, is number two. Investment has an impact on both costs and assets; that is, it's associated with pluses and minuses on the balance sheet. Investment is normally in third place by quite a bit, because it represents a tradeoff between more value and more money tied up.

When trying to improve results by cutting operating expenses, we end up looking everywhere for places to improve. The project manager spends much of her time searching for ways to use fewer resources. The lack of focus should warn us that we may not be dealing with a methodology that will lead to dramatic improvements. We should be suspicious. So next let's work out a concrete example, and compare the importance of operating expense and throughput.

Cost Vs. Throughput

You are the CEO of a fairly large company that develops software for other companies. If someone wants some software developed, but doesn't want to make the commitment to hire the necessary development group, they might call your company. You don't need to know a lot about the technical side of the business, and in fact you have risen to your current post because of your sales abilities rather than your technical acumen. This hundred-million-dollar company has a simplified balance sheet that looks something like that in Table 15-1 for the most recent year (all figures in millions of dollars).[7]

Table 15-1   Balance Sheet

	Current
Sales	$100.0
Materials	5.0
Throughput	**95.0**
Direct labor	30.0
Overhead	60.0
Operating Expense	**90.0**
Net profit	**$ 5.0**

First, make sure the numbers are clear. Throughput equals sales minus (truly variable) materials; operating expense includes direct labor and overhead. In order to simplify the example, we haven't included any investment.

Your company is doing all right so far. You're not losing money, although the board would naturally like to see you make a bigger profit. Based on discussions with various high-level managers and soothsayers, you've come to the conclusion that you have 20% excess capacity. That means, for example, that direct labor costs can be cut by 20% with no adverse effects on existing or forecast business. Assume this is true. Assume also that your current market is saturated. The market is price sensitive, but you'd have to sell at a 20% markdown — way below cost — to get much additional business. What should your actions be? What will be the bottom line results? Please think about your answer for a few moments, work out the numbers, and then read on.

The typical corporate reaction will be first to evaluate the effect of cutting costs. Let's adjust the numbers, decreasing direct labor costs by 20%, and look at the results (Table 15-2). This seems to be a real improvement and a good decision. By cutting direct labor costs, net profit is up by 120%. You might have some morale problems, but frankly, cost cutting is standard operating procedure these days. The remaining people should just be glad they still have their jobs. We live in a difficult and highly competitive business climate. With a profitable move like this, at least your job is secure.

What would happen if someone suggested that you should use this extra capacity to sell more projects, instead of laying people off? You know that the market is saturated. Based on the balance sheet, total cost averages about 95% of selling price; with the required 20% markdown to get more sales, you'd be selling at 80% of the normal selling price, or about 15% below cost.

Table 15-2 Balance Sheet (Planned Cost Cutting)

	Current	Cut OE
Sales	$100.0	$100.0
Materials	5.0	5.0
Throughput	**95.0**	**95.0**
Direct labor	30.0	24.0
Overhead	60.0	60.0
Operating Expense	**90.0**	**84.0**
Net profit	**$ 5.0**	**11.0**

Table 15-3   Increase Throughput (1)

	Current	Cut OE	Raise T
Sales	$100.0	$100.0	$116.0
Materials	5.0	5.0	6.0
Throughput	**95.0**	**95.0**	**110.0**
Direct labor	30.0	24.0	30.0
Overhead	60.0	60.0	60.0
Operating Expense	**90.0**	**84.0**	**90.0**
Net profit	**$ 5.0**	**11.0**	**20.0**

Table 15-4   Increase Throughput (2)

	Current	Cut OE	Raise T
Sales	$100.0	$100.0	$120.00
Materials	5.0	5.0	6.25
Throughput	**95.0**	**95.0**	**113.75**
Direct labor	30.0	24.0	30.00
Overhead	60.0	60.0	60.00
Operating Expense	**90.0**	**84.0**	**90.00**
Net profit	**$ 5.0**	**11.0**	**23.75**

It sounds crazy, and you've already got a perfectly good solution. But for the sake of completeness let's work out these numbers.

You could sell $120 million in projects, but you'd have to cut 20% off the extra $20 million for the discount, making our total additional sales $16 million. Material costs would also go up 20%, and based on our extra capacity labor costs would stay about the same (Table 15-3). By choosing to increase throughput rather than cut costs, the net profit has increased by a factor of four instead of just over two.

In fact, there's an error in these calculations that ignores another important aspect of throughput. If only half our capacity is producing, we can make twice as much, not 50% more. If we're 80% (4/5) productive, we can produce 5/4 as much, or 25% more. Factoring in the 20% discount, the numbers in the table should read as shown in Table 15-4.

We've made a few assumptions here that should be brought up, although they don't affect the conclusions significantly. First, we've assumed that overhead costs couldn't be brought down along with costs of direct labor. This is typically the case. We've also assumed that our new customers to whom we are giving the discount won't tell our old customers about it, thus forcing us to sell at lower prices to everyone. This kind of "market segmentation" isn't so difficult in the software contracting business, but in many markets it requires careful planning.[8]

Standard cost accounting would not even consider the option of generating more throughput in this situation, because you would have to sell below cost. Yet when looked at in a global context it is extremely logical. Realistically, what's the limit on cutting costs? You can cut them down to zero, if you lay everyone off. This happens, especially over the last few years with defense contractors. On the other hand, what's the limit on increasing throughput? Unless you own your entire market, there is none; and even then you can expand into other markets. Throughput is unlimited; this is a very important message.

Perhaps it makes sense to consider switching the priorities of throughput and operating expense, and making throughput number one. This is not common practice. Does it seems reasonable that such a fundamental shift in priorities could be valid, and no one would have thought of it?

Automobile Production

Many Japanese companies have been throughput-oriented for years. The most visible component of this for those who buy Japanese goods has been the improvement in their quality. Consider the change in U.S. consumer perception of Japanese manufactured goods between the 1950s and today. In the 1950s Japanese products were considered to be junk. "Made in Japan" was a synonym for "cheap." Today many Japanese products are considered the best in the world. Which compact car now costs more, a Ford or a Toyota? Which has a better reputation for quality?

There are lower costs associated with higher quality — less wasted material, lower warranty costs, and so on. In many kinds of project work, these costs are even more significant because higher quality implies less need to go back and redo earlier work. Nevertheless, for both manufacturing and projects those lower costs are dwarfed by the increases in throughput that are also associated with higher quality. Consider Figure 15-1, comparing passenger car production of the U.S. and Japan over the last 40 years.

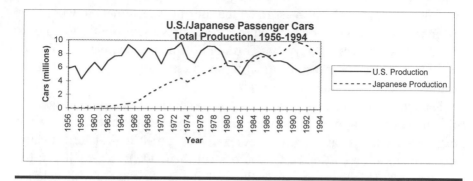

Figure 15-1   U.S./Japanese Passenger Car Production.[9]

The steady increase of Japanese production is quite striking, especially when compared with the cyclical nature of U.S. production. However, it does look as though the balance has shifted in the last few years. Before we draw that conclusion, let's make sure we're talking about the right systems. Are we talking about the countries of the United States and Japan, or are we talking about U.S. and Japanese auto makers? If the latter, we have to account for the fact that many Japanese companies have been assembling automobiles in the U.S. over the last few years. If we shift the automobiles assembled in the U.S. by Honda, Toyota, Nissan, Mazda, and Subaru over to the Japanese side, we get a slightly different picture (Figure 15-2).

Figure 15-2 is probably more indicative of the relative progress of Japanese and U.S. auto makers. The steady increase of Japanese car production implies a constant focus on increasing throughput. They've been a step ahead the whole way. There is also a related capability for lifetime employment. How many layoffs do you think were required in the Japanese auto industry during this period? It's interesting to think of lifetime employment as a capitalist concept.

Figure 15-3 shows the growth of passenger car exports from the U.S. and Japan. Japanese companies, with a solid base of sales at home, can afford to sell their products — meaning their extra capacity — around the world, without having to worry about "making their margins." If their local markets are sewn up and provide a dependable revenue stream, any products that can be sold abroad for prices above material costs will have a positive impact on the bottom line. The foreign markets can be used as kind of a demand "reservoir" to protect capacity and workers against market fluctuations. This happens to such an extent that there are strong U.S. trade laws in effect against foreign suppliers "dumping," or selling below "cost."

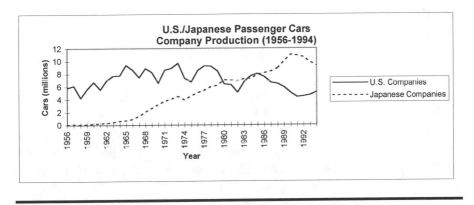

Figure 15-2   U.S./Japanese Passenger Car Exports.[10]

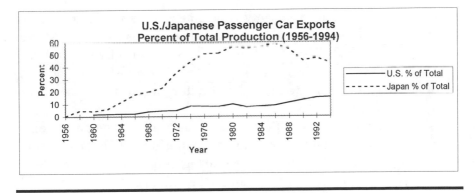

Figure 15-3   U.S./Japanese Passenger Car Exports.[11]

This is a manufacturing example, but nothing prevents construction companies, software companies, or anyone else from doing the same thing. Product costing is wrong. Excess capacity can be sold cheaply and still make money.

Historically the competitiveness of the Japanese has been attributed to cheap products, cheap labor, or use of robots to lower production costs. As their products have become high quality, as their labor force has become more highly paid, and as the use of robots have proliferated in other parts of the world, these excuses have been exposed as erroneous.

Total Quality

Current wisdom tends to focus on the Total Quality Management movement (TQM), which has been active in Japan since the late 1940s. TQM stresses the importance of quality in all aspects of a company; not just in the products themselves, but with everything that takes place. TQM really took off in Japan in 1950 when Dr. W. Edwards Deming, one of the founders of the quality movement, came to Japan and gave lectures on quality control. Since then the quality of Japanese products has risen consistently. The quality movement in Japan is clearly a key part of the Japanese competitiveness. How does it relate to throughput?

Generally speaking, the market always has long-term importance in gaining future sales for any company. In TOC terms, the market is always, at least strategically, a leverage point. Quality is about meeting and exceeding market requirements. In most markets, high-quality products are highly competitive products. A focus on quality translates to a focus on throughput.

Furthermore, a focus on quality is definitely a focus away from costs. From a cost-oriented point of view, the cost of really satisfying one customer will often be too high. The cost of improving processes may be exorbitant compared with the immediate benefits. And yet still the customer- and quality-oriented approach makes sense, because the customers are paying the bills, now and in the future.

Is a focus on quality always the best way to increase throughput? In other words, is it always leverage point? Short-term, quality will not be a leverage point if the customers don't know they need higher quality. However, as soon as a company has higher-quality products, they start positioning themselves as a quality producer. This enables them to sell more, possibly at higher prices. Once the perception of need for higher quality becomes established in the market, quality becomes a necessary condition without which the game can't be played. American automobile manufacturers went through major quality upgrades in the 1970s and 1980s just in order to be able to play the game.

Long-term, throughput is unlimited. The way to get more throughput is through the market, which means the market is always a strategic leverage point. That means quality will remain valuable insofar as it helps people to focus on throughput and the market. We can restate this conclusion by saying quality is of long-term value if it pushes people to look outside the box of their current system. This "out-of-the-box" thinking is the subject of the next chapter.

Key Concepts

- Any additional throughput made from spare capacity goes straight to the bottom line. This can provide a great deal of leverage in getting more sales.
- Throughput is unlimited.
- Throughput is number one, *"ichiban."*
- Japanese companies have been throughput-oriented for many years.
- The market is key in getting more throughput.
- Quality is often key in getting more market.

Questions for Further Thought

1. Obtain a balance sheet and work out the examples from Tables 15-1 to 15-4 for a real company. Does throughput still look like a better leverage point than costs?
2. Will a focus on quality cause you to come up with the better answer in the example of Tables 15-1 to 15-4? What about a focus on lifetime employment?

16 The Throughput World: Climbing Out of the Box

In the previous chapter we stressed the importance of throughput. There is a term for the paradigm that stresses a strategic emphasis on throughput: the **Throughput World**.[12] But Throughput World thinking requires much more than just an emphasis on throughput. It requires a system-level viewpoint, which means looking at the entire system rather than the individual parts. It requires an understanding of constraints or leverage points, which are key areas to focus on for gaining improvements. We will deal with these concepts in this chapter.

The Systems-Level Viewpoint

The system in Figure 14-1 is surrounded by a box. That box differentiates the system from the rest of the world. The pieces inside the box must work together to generate throughput; otherwise we wouldn't include them in the system. Nevertheless, the boxes around systems are mainly constructs of our minds. A box may or may not have any tangible existence. Systems change over time; our definitions of systems change. Children grow up and get married, people are born and die, families move. Companies are bought and sold; they have changing chains of suppliers, customers, and employees. A **system-level viewpoint** implies the following:

- A recognition that there are distinct systems
- An understanding that we have defined the boundaries of the box, and that therefore we can change those boundaries, and
- A realization that the parts of a system must work together toward the goals of the whole.

A preoccupation with costs keeps attention inside the box of the current system. It keeps people looking everywhere for improvement. The sky is outside the box, and that's the limit of throughput. The shift in thinking from reducing costs to increasing throughput is called moving from the **Cost World** to the **Throughput World**.[13] As we saw from the example in Tables 15-1 to 15-4, concentrating on throughput can be much more powerful than concentrating on costs.

By the way, don't confuse the importance of *focusing* on throughput instead of costs, with the importance of the throughput and costs themselves. In bottom-line measures such as net profit and return-on-investment, operating expense and investment can have a bigger impact on the bottom line than throughput. The Throughput World implies a strategic emphasis on throughput, which in the long run is likely to produce much bigger gains than the traditional emphasis on costs.

Consider how the focus of our manufacturing and service organizations has changed over the last 20 years. More and more the "customer-oriented" business is becoming the norm. Consumer perception of quality and service is a huge competitive advantage. "Quality is Job 1." "We're Ready when You Are." This is a manifestation of the "out-of-the-box" direction that the world is moving in. This direction is largely due to the contributions and successes of the Japanese in recognizing the importance of throughput.

Example: DOD Contracting

In Chapter 2, Janet made the following statement about the bid she was preparing:

> It's a shame that we only get a tenth of the bids we make, but somehow our costs are just too high. Our cost structure really cuts into our competitiveness; we're constantly losing out due to price. It's not that management hasn't been trying to downsize; counting losses of production people, we have only a third the number of employees we had ten years ago. That's how it is in the military contracting business these days; you make drastic cuts just to keep the company afloat. But now we're stuck with even higher overhead ratios. It's a spiral that's sucked a lot of companies down, and we're starting to see the bottom of the tub ourselves.

Traditionally we look at a combination of product cost, selling price, and margin to decide whether or not to bid on a project. We fiddle around with

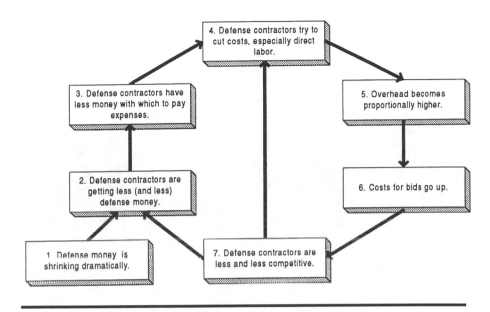

Figure 16-1   Defense Death Spiral

the numbers inside the box. As part of the process we allocate all the overhead costs, from administration to zambonis. So longer-term, we'd always like to cut costs in order to bring down product costs and be able to sell at more competitive prices. Let's try to verbalize more precisely the downward spiral Janet described.

Figure 16-1 is a Current Reality Tree, similar in structure to the one we saw in Chapter 5. This example is very small, and is used here as a means of communicating a typical problem and its causes. The instructions for reading are the same as those for Figure 5-1. Start at the bottom left with the box having no arrows going into it — box number 1. Read each box in order, checking to make sure that it makes sense. Read the arrows "if ⟨starting entity⟩ **then** ⟨ending entity⟩" to check the causalities. Read all arrows, including the one between boxes 7 and 4. For example, start by reading "**if** defense money is shrinking dramatically, **then** defense contractors are getting less (and less) defense money." The "(and less)" in box 2 is a convention that implies a loop. This diagram has a couple of important loops. As costs are cut, overheads become a larger proportion of costs, and therefore the prices which must be quoted on bids go up. There's another loop triggered by the first, around boxes 4, 5, 6, and 7: the Cost World mentality. If the standard method of improving the bottom line is through cost cutting, this loop will continue.

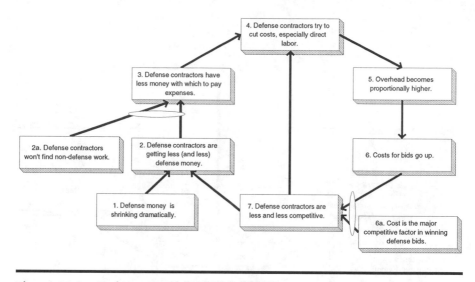

Figure 16-2   Defense Death Spiral (revised)

Given box 1 and the common focus on costs, this diagram looks inevitable. It's no wonder people feel a need to add something more to the cost accounting decision process. In fact, there are many ways people break out of these loops. Every arrow in the diagram has assumptions behind it, many of these assumptions that can be challenged. For example, direct labor cuts can be matched with overhead cuts. This process of downsizing (or, more politically correct, "rightsizing") often implies a complete re-engineering of the business. Is cost cutting the best approach?

Let's look at Figure 16-1 with a system-level viewpoint, and try to see more clearly the box we're in and some of the assumptions behind it. One assumption is that only defense work can be done by DOD companies. In fact, some DOD companies have branched into nondefense work with great success. Another assumption is that the only way to become more competitive in the bidding process is through lower costs. In reality, as we've seen with the $100 million company example from Chapter 15, using extra capacity to generate more throughput can be extremely powerful if you're willing to consider it. In some cases being able to deliver quickly and reliably is also important. We can add some of these assumptions to Figure 16-1 to get a more useful picture, Figure 16-2.

As we saw in Chapter 5, the ellipse binds together arrows. An ellipse is read as "and," meaning all causes joined at the bases of the arrows encircled by ellipses are required to achieve the effect at the head. Reading 2 and 2a to

3 would give: "**if** companies get fewer (and fewer) contracts **and** DOD contractors won't find non-DOD work, **then** companies have less money with which to pay expenses."

These "additional causes" are very important. They verbalize assumptions being made about the system, assumptions that can be challenged. Many companies choose not to challenge these assumptions and continue down the spiral. Box 2a suggests one alternative approach: finding nondefense work. Box 6a suggests another slightly more subtle approach: not relying on costs when making bids. In the $100 million company example from Chapter 15, additional products were sold below cost, because the additional throughput went directly to the bottom line.

Constraints and Leverage Points

The system-level viewpoint means we will keep looking to make real, signif icant improvements to the whole system or organization. We must keep our eye on the ball, which is the bottom line. The "out-of-the-box" mindset implies that we're not just looking inside the organization to find improvements.

There's a simple situation that illustrates the importance of looking out of the box. In the development of new products, the difference between being first to get a product to market and being second can be huge. In consumer electronics, for example, it might mean thousands of dollars in extra profit for each day that a development project finishes early. Nevertheless, this tradeoff is often not evaluated. The costs of resources required to achieve the improvements might be considered; the impact of using those resources is not. Highly leveraged means of shortening projects, for example by hiring temporary clerical help or by offering employee incentives, are often not considered. We keep looking inside the box.

In some ways the need to look out of the box makes the challenges of improving seem even more daunting. It's a big world, and chances are that very few of the places we might try to improve would have a significant positive impact on the bottom line. In order avoid local optima and meaningless improvements we need to evaluate the global impact of actions. Actions having a global impact always come with many interdependencies, which implies bigger risks. Improving performance for an entire project seems much more difficult than improving performance for a single task.

Still, we would like to believe that some key improvement points exist and that we can identify them. These key points are called **constraints**. A constraint is anything that limits our ability to improve the system. Constraints are also often called **leverage points**, because that term points out that your limitations imply opportunities to improve.[14] We sometimes use the term **throughput lever** when the leverage points are Throughput World oriented; most of the big ones are. A leverage point is anything that enables us to improve; a throughput lever is a leverage point that is focused on throughput.

One might become concerned that there are too many leverage points, rather than too few. This is an understandable concern that may be part of a residual Cost World mentality. In the Cost World, you need to look everywhere in the box. Costs can be cut anywhere, improvements in efficiency can be made anywhere. Of course, this carries with it an interesting problem: all such improvements tend to have a relatively small impact on the bottom line. There are many tradeoffs required when deciding how and where to improve. Evaluating the tradeoffs can be a formidable job.

When we think of a few leverage points allowing us the most improvement, sometimes we think of the Pareto 80/20 rule (technically a type of "power law"[15]), which indicates that about 80% of the improvement comes from addressing 20% of the leverage points. Right away we know that we shouldn't be trying to improve everywhere. But before we start looking, there's some bad news and good news to think about. The bad news is that even having to look at 20% of everything is a lot to look at.

The good news is that the Pareto principle deals with related but independent events. Action B isn't directly influenced by action A. In most organizations we have chains of dependent events. In order to carry out action B, first action A must take place. Throughput is produced by actions of the entire chain, rather than individual links. So throughput is governed by the constraints (weakest links) or leverage points, and the number of weakest links tends to be very small. Instead of 20% of a set of independent leverage points having 80% of the impact, we more likely have 1% of a set of interdependent leverage points producing 95% of the impact.

An important implication of the 95/1 rule (rather than 80/20) is that not everything in a system needs attention. If we can achieve significant improvements by focusing efforts in a very few places, there's clearly not a need to change everything. This means that, while TOC may imply a radical shift in our thinking, it also implies a measured approach to our actions that is much less destabilizing than many current re-engineering and "rightsizing" approaches. In fact, TOC techniques have been used frequently in the manufacturing world to focus more technically oriented improvement techniques associated with Total Quality Management (TQM) and Just-In-Time (JIT).

There are numerous examples of leverage points given throughout this book. In Part II, we saw that the Critical Chain can be a leverage point. In Chapter 15, we saw that throughput itself can be highly leveraged. The Current Reality Trees in Chapters 5 and 6 also help identify leverage points, namely core problems. Leverage points do exist, but you have to look for them. If you don't believe they exist, you won't look for them, and they will have to find you. This is not an uncommon management style; neither is it a promising one. Looking at it another way, the only thing that will be lost by assuming there are leverage points is some time you could be spending fighting fires. But typically those fires are caused by not addressing the leverage points.

In Figure 16-1, the obvious "core problem" is that defense money is shrinking dramatically. Contractors naturally wish there were more defense money. How can they know that's not a leverage point? Because it's outside most companies' control. We have to believe there are leverage points that we can influence. Wishful thinking is not a leverage point. To quote the popular saying, "if the horse you are riding dies, get off."

"Optimization" — The Best?

A concept closely allied with "in-the-box" thinking is "optimization." It seems as if, given the choice, it would always be desirable to optimize a system or the parts of it. That means doing the best with what we have; it's common sense. Yet if we consider this at a philosophical level, it's clear that optimization implies limits. You can't optimize a system with no boundaries. We'll typically expect the return on optimization to be a few percent lower costs or higher productivity. This return is inherently limited by the boundaries you have placed on the box you're optimizing, which means that optimization can't be part of a long-term improvement process. Furthermore, when the system (including possibly its relationships with other systems) changes significantly, as it must, it needs to be reoptimized. It doesn't look as if optimization is a useful long-term strategy.

In reality, "optimization" is not just a limited strategy; it can actually be harmful, even in the short term, for two reasons. First, people's attention is directed the wrong way. To optimize they must be strongly focused into the box. This promotes a culture that can make big increases in throughput very difficult.

Second, the actions required to produce local optima aren't necessarily the actions needed to produce global improvements. In other words, the sum of the local optima doesn't necessarily equal the global optimum. This still may not sound terrible; we might be tempted to argue that a local optimum, while not helping the global picture, might not necessarily be bad. It certainly keeps people feeling useful.

The reality is that any organization has limited resources. Those resources must be allocated to different work groups and jobs based on what seems to be best for the entire organization. Allocation of resources where they will produce a local optimum, and where the global results are not significantly affected, is more than a waste of the associated costs; it is a loss of those resources for the entire system. Those resources can be money, time, and the good will of stakeholders. This means that local optimization can indeed cause global performance to go down.

This still doesn't sound so bad if one comes at it with a Cost World mentality. Even taking into account the Pareto principle, the Cost World point of view assumes that around 20% of the places we look will help us to get real improvements. But if we add the reality of leverage points, the reality that 80/20 becomes 95/1 in systems of dependent events, we must come to the conclusion that "improvements" just to support local optima are a very bad idea. Consider the following statements from the manufacturing world, keeping in mind that in many cases limited capacity or "bottleneck" resources are in fact leverage points:

- "An hour lost at a bottleneck is an hour lost for the total system."[16]
- "An hour saved at a nonbottleneck is just a mirage."[17]
- "Motto: The sum of the local optimums is not equal to the global optimum."[18]

Let's take a concrete example. U.S. Air Force repair and overhaul depots have a discrete function for matériel transport. This function has historically been measured based on costs, such as cost-per-mile. The lower the costs for transport, the better. This provided a strong incentive for the transport function to send things as cheaply as possible. This generally meant they wanted to send things as slowly as possible.[19] In-house transport has also needed to be efficient. There have been, for example, incentives to wait until a truck was full before shipping its contents. Sending things more quickly required permission for the additional expenses, which resulted in queues of items that couldn't be shipped until the decisions were made. Those who needed parts had to wait significant time — often weeks more than necessary — before they could get them.

These transportation policies had a significant impact on the time required to repair multi-million-dollar aircraft. The increased transportation times meant that airplanes would stay in the repair and overhaul system much longer than necessary. This in turn required either a corresponding increase in the number of planes in the Air Force system, meaning increased investment, or a decrease in the number of mission-ready planes, meaning a decrease in defense potential or throughput. The cost savings for being "efficient" were negligible, especially compared with the global impact on investment and throughput. In fact, there were significant costs associated with the paperwork and approval process, to the extent that it wasn't clear that even a local optimum was being achieved. In any event, the global measurements clearly indicate that the local optimum was globally suboptimal. The U.S. Air Force has already made important TOC-related improvements, especially to the matériel transport system, under the banner "Lean Logistics."[20]

Not-for-Profit Throughput

As mentioned in Chapter 14, throughput is difficult to measure in a not-for-profit environment. Of what value is it to have such a measurement? Is it worth the effort of creating it? Let's consider a concrete situation. Suppose there as a measurement of the throughput of health care for which associated ethical, moral, political, and technical issues were resolved — the "super" quality-adjusted life year, or SQALY (refer to Chapter 14 for the definition of a QALY). Suppose also that the money that can be spent on health care is limited, forming a "necessary condition" that cannot be violated. The obvious use of the SQALY would be to determine cost per unit of throughput for different types of illnesses and treatments. Note that this is just the reciprocal of the Throughput/Operating Expense productivity measurement mentioned in Chapter 14. Health care rationing could then be done on a "rational" basis, by ranking services and providing only those that give the most throughput per dollar spent. Of course, if (for example) the appendectomy were discovered to give the best use of health care per dollar, it would be ludicrous to put all available resources into appendectomies. The existing market demand must clearly be taken into account when deciding how far down the list of services to go. This type of QALY approach has in fact been taken in the state of Oregon.[21]

Suppose now that we also know the approximate demand for different services, so we know what will be required of the health care system. Do we have the complete picture? Let's say, hypothetically, that major organ transplants fall to the bottom of the list. They are expensive and high-risk, and therefore don't generate enough SQALYs per dollar. Suppose, further, that there are existing health-care facilities that have qualified personnel and special, expensive equipment to perform these operations. Given that the equipment and expertise exist, the real marginal cost of performing the operations might be relatively small. The fully allocated cost that shows up in the rationing system will be much higher. Existing capacity has not been taken into account. We have to question how effectively looking at the cost element of the equation by itself is going to allow us to maximize performance of the system, when it doesn't take into account the ability to generate throughput.

The lesson is not that throughput is useless for measuring health care, or that we somehow need "better" product costs. We don't want to imply that Oregon's approach is a step backward. The real Throughput World message is that the providers of health care should be trying to produce as much real throughput as possible for the money spent. This is their goal, and its throughput could be defined by the SQALY. Maximizing SQALY production would seem to be a good goal for a national health care system.

Imagine that some of the rewards for health care management employees were determined by throughput generated per operating expense dollar spent at the organization level — Throughput/Operating Expense, in this case SQALY per dollar. Imagine, further, that everyone understood the fundamental Throughput World concepts: the unlimited potential for throughput generation, the true marginal costs required to produce more throughput, the systems-level viewpoint, and so on. What would be the impact? Health care organizations might consider more carefully the effects of demand and capacity on the global measurements. They would be more likely to try and improve the global system. The emphasis shifts from how to ration care, meaning how to control costs, to how to produce more for the money we're spending. Probably, in the end, the details of the SQALY construction would be much less important than the Throughput World mentality on improving the overall system.

This example is not intended as a "solution" to any existing health care problems. There are far too many political and technical obstacles to attempt that here. It is intended to show the value of throughput as a measurement, and to show how Throughput World thinking can give a different, and perhaps useful, perspective on a chronic problem.

The Japanese Advantage

Let's briefly go back to the previous chapter, and consider another reason attributed to the Japanese for their competitive advantages. Michael Rothschild argues that Japanese organizations have an ability to accumulate experience more quickly than their competitors.[22] They learn more quickly. They can incorporate new concepts more quickly. They are always ahead.

This is clearly a useful attribute, and it is not in conflict with the Throughput World. In fact, the ability of an organization to accumulate experience can be considered an effect of the Throughput World. The Throughput World requires looking out of the box. It requires viewing the organization as a chain. It requires cooperation of the entire system. This same viewpoint, understood widely enough and deeply enough throughout the organization, also provides a motivation for the kind of cooperation that makes "organizational learning" possible.

The Throughput World concepts are useful by themselves, and certainly help to point out many differences between common practice and common sense. In order to make them practical we need more understanding still. We need a process for managing leverage points. We need to know how to apply these concepts in day-to-day decision making. We need to understand what kinds of local measurements are needed to promote Throughput World thinking among people within organizations. The next chapter describes a five-step process for managing leverage points. Chapter 18 describes some basic aspects of TOC decision making. Discussions of measurements are postponed until Part IV because they relate closely to the implementation process.

Key Concepts

- Leverage points exist.
- In most organizations the 80/20 rule is actually closer to 95/1.
- The sum of the local optima does not equal the global optimum.
- Throughput is more useful for determining how to increase throughput than for determining how to decrease operating expense.

Questions for Further Thought

1. Sun-Tzu, arguably the greatest writer of all time on the subject of warfare, said that his enemy must not know where he wants to give battle. He wrote of his opponent: "For if [the opponent] prepares to the front his rear will be weak, and if to the rear, his front will be fragile. If he prepares to the left, his right will be vulnerable and if to the right, there will be few on his left. And when he prepares everywhere he will be weak everywhere."[23] Do you think Sun-Tzu advocated a Cost World or a Throughput World approach to warfare? Where should his enemy prepare?

2. There are important assumptions in Figure 16-1 that haven't yet been surfaced. Try to come up with some.

3. "Best practice" is a popular concept that involves learning and applying those ideas and techniques that have been proven to work well for others. What concept discussed in this chapter is "best practice" often similar to? What kinds of questions should you ask about "best practices?"

4. There is a potential drawback to using throughput per operating expense dollar (T/OE) as a measurement. What is it? How would you get around this drawback?

17 Global Improvement: The Five Focusing Steps

In order to continue to make improvements over time, we need a process. We need a set of steps to help us focus our efforts on the leverage points. Such a process was discussed briefly in Chapter 8; let's now take it further.

We'll look at a small company that builds custom hardware devices for computers. This company has one customer service representative, one design engineer, one technician, and one computer programmer. For each project, the customer service representative will typically spend about 1.5 weeks on-site to work out the requirements with the customer. Then the design engineer will spend about the same amount of time working out detailed specifications. When the specifications are ready, the technician builds the hardware (1 week) while the programmer writes the needed programs (3 weeks). After 2 more weeks for both the programmer and the technician to integrate and test, the final product is ready to be shipped. The flow of work looks like that shown in Figure 17-1.

On average, how many projects can be completed in a year, assuming 50 working weeks? The person spending the most time on each project is the programmer; it takes him 5 weeks per project, compared with 3 for the technician and 1.5 each for the customer services rep and the design engineer. The theoretical maximum number of projects, assuming we start with some work in the system, is ten; the programmer clearly can't do more. Where is the throughput lever?

If you answered "programming," you spoke too quickly. If there were demand for less than ten projects per year, the throughput lever would be in the market. In order to get more throughput, more sales would be needed. But let's assume for the moment that the demand is (again, on average) for

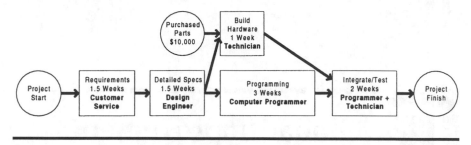

Figure 17-1   Custom Hardware Project

18 projects per year, so the programming resource really is the throughput lever. An increase in capacity in programming will directly improve the bottom line of the entire company.

It would be logical to focus on programming and try to improve that area first. Let's assume that in fact an improvement is worked out, so that the programmer is now spending half as long — 2.5 weeks — per project. This is not an unreasonable assumption. Typically major improvements can be made in individual work centers when attention is concentrated there. For example, the programmer should not be answering phones, filling out unnecessary reports, or playing too many computer games.

Does that improvement all result in an improvement of the bottom line? No, because next the technician — who spends a total of 3 weeks per project — will become the constraint, and hence the throughput lever. And if next the technician constraint is broken, so that the technician is spending only 2 weeks per project, where is the throughput lever?[24]

Going back to the original Figure 17-1, with a market demand of 18 projects per year, suppose again the programmer is the leverage point. What is the meaning of the programmer being idle? Quite simply, every minute of time the programmer isn't working is a minute of lost throughput for the entire company. This means that other resources should be trying their best to make sure the programmer isn't idle. They should make sure that work is waiting. Of course, they also don't want to produce too much work, or delays (work in process) will build up in the system and lead times will suffer.

What is the meaning of the design engineer being idle? The answer is, we expect it. His capacity is greater than demand, so he will more than keep up with the programmer. We certainly don't want him to overproduce.

Finally, if despite all our efforts the leverage point remained in programming, we still have a choice: we can hire another programmer, and thereby "elevate" the throughput lever. Of course, as happened above, the lever will then be elsewhere.

The exercise we're going through with these projects suggests the five-step improvement process of TOC:[25]

1. **Identify** the leverage point(s).
2. **Exploit** the leverage point(s).
3. **Subordinate** everything else to the above decisions.
4. **Elevate** the leverage point(s).
5. Go back to Step 1; don't let **inertia** become a constraint.

We first identified the programmer as the leverage point. Next, Step 2, we probably tried everything possible to exploit the leverage point, or squeeze the most possible out of it. For example, the programmer might have been relieved of answering the phone or cleaning the coffee machine. He might even be given incentives to produce more.

The third step, "subordinate," really means the subordination of everyone to the global objectives. Everyone should be contributing to the goal. Since the programmer's activity directly relates to the bottom line, everyone is trying to keep him productive (without overproducing, which would needlessly build up work in process). This is somewhat different from the standard meaning of subordinate. People are not subordinate to each other; as part of the organization, they subordinate themselves to the organization's global objectives.[26] This is an uncommon point of view in cost-oriented companies, because widespread understanding of the global objectives and leverage points is uncommon.

The fourth step requires getting still more of the leverage point; it also implies a significant increase in operating expense. For example, overtime or hiring a new worker are definitely considered elevation. Step 4 consequently tends to have long-term strategic implications.

Throughout the whole process, the situation must be constantly reevaluated. If the location of a leverage point changes, everyone's behavior will need to change. For example, suppose the technician suddenly quit and a new, poorly trained person took over. At least temporarily the technician might become the leverage point. In that situation it might make sense to allow the programmer to answer the phone. While we put in this reevaluation as Step 5, in reality it must be performed constantly.

The Five-Step Logic Tree

The five steps can be viewed as a logic tree, as shown in Figure 17-2. This tree is known as a "Future Reality Tree" because it's not assumed to be our

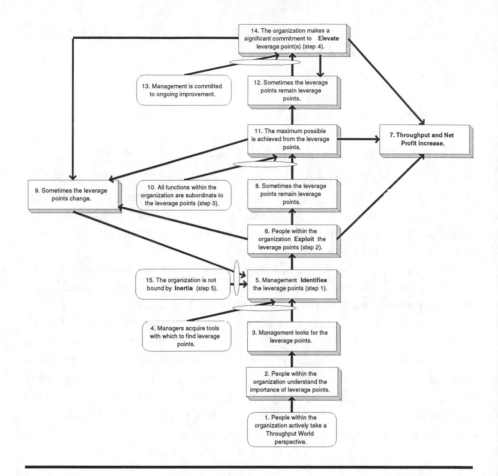

Figure 17-2   The Five-Step Improvement Process

current reality. There are consequently a few differences from the previous Current Reality Trees. Certain changes are needed to our current reality to cause the higher-level effects of this tree to happen. These changes are called **injections**, because we assume they're injected into the current reality without worrying about how.[27] We will worry about how to accomplish them after we decide we really want them. The injections in this diagram are inside the rounded boxes. This tree should be read in the same way as Figures 5-1, 6-1, and 16-1, with the ellipses binding together the arrows and being read as "and." The only difference is that the rounded boxes aren't required to be part of our current reality.

It's important to observe that when we accept an entity as valid, that entity is assumed to be valid for all entities it leads to. For example, box 6 is an effect of box 5, but it also assumes that box 2 exists.

We start with the injection "people within the organization actively take a Throughput World perspective"(1). In fact, this tree can be taken as defining many of the implications of a Throughput World perspective. If people are taking this viewpoint, by definition they must understand the importance of leverage points (2). Of course, understanding the importance of leverage points implies that management will be looking for leverage points (3).

It's not always obvious where the leverage points are, so we need another injection: managers acquire tools with which to find leverage points (4). The tools might be conceptual tools, such as the ability to create a Current Reality Tree; they might also be software that helps to identify and manage Critical Chains. Of course, if management looks for leverage points and acquires the tools, they will identify the leverage points (5), which is Step 1 of the process.

If management is able to identify the leverage points, people within the organization (having from box 1 a Throughput World perspective) will carry out Step 2 and exploit the leverage points (6). This by itself can lead to increases in throughput (7). In addition, either the leverage points will remain in the same places (8), or they will change (9).

Let's first assume some leverage points remain the same. If we add Step 3 of the process as another injection, that all functions within the organization are subordinate to the leverage points (10), then we will be achieving the maximum possible from the leverage points (11). It's important that box 10 is an injection; this kind of subordination isn't necessarily simple to achieve. Because we're getting the maximum possible from the leverage points, throughput again goes up (7). Furthermore, once again either the leverage points will remain the same (12) or they will change (9).

Now we have another injection, that management is committed to ongoing improvement (13). This means global improvement, as implied by box 1. It also means not just paying lip service to global improvement, especially since there are frequently many pressures at every level of management to achieve local improvements. Given this commitment to further improvement, we can confidently expect that Step 4, elevation of the leverage points, will happen (14). Once again throughput increases (7), and the leverage points will either stay the same (12) or change (9).

We can finally complete the loop that makes the process repeatable, through an injection that the organization is not bound by inertia (15). Assuming that the leverage points change, and the organization is not bound by inertia, the new leverage points will be identified (5) and the cycle can continue. This loop of identifying/breaking/reidentifying leverage points, shown between boxes 5, 6 (or 11 or 14), and 9, implies that the process will continue to produce more and more positive effects. The main positive effect shown is an increase in throughput and net profit, box 7.

There's an important point in this tree that isn't obvious from the original five steps. The arrow from box 14 to box 12 implies that elevation of the leverage points may not result in a new leverage point. Of course, that's fine as long as throughput continues to go up, but there's more than that. Often it's not desirable for the current leverage point to change.

Longer-Term Success: The Revised Five Steps

When a company starts to make the shift to the Throughput World, typically they will follow this five-step process. They will identify leverage points and then break them, sometimes fairly quickly. There are two problems with this process. First, a frequent shift in leverage points implies a frequent shift in focus for everyone. The focal point of the company is changing, sometimes arbitrarily. By this time many people, including sales and marketing, as well as any product development and production areas, should be involved in exploiting and subordinating to the leverage points. Given the management attention on the current leverage points, given their importance in running the business, this kind of shift in focus can be dangerously destabilizing. People will start to feel like they're watching a tennis match as the leverage point bounces back and forth.

Second, during this process there can come a point where the throughput lever is a resource that is very difficult or perhaps undesirable to elevate. It may be a place that requires large capital investment; it may be a proprietary technology or concept that distinguishes the company from the competition; it may just be a control point that is easy for other resources to subordinate to, making easier the management of the entire organization.

These two problems suggest that perhaps management can take control. Rather than allowing leverage points to wander around, they should select where the leverage points should be. We'd prefer to balance improvements in the leverage points with improvements elsewhere so that the leverage points stay in the same place. When that happens, the five steps change somewhat:

1. **Select** the leverage point(s).
2. **Exploit** the leverage point(s).
3. **Subordinate** everything else to the above decisions.
4. **Elevate** the leverage point(s).
5. Before making any significant changes, **evaluate** whether the leverage point(s) will and should stay the same.

We want to be proactive in controlling where the leverage points are and where the focus of the organization is. This doesn't mean we stop improving; it means we control the improvement process much better. The implications of this change are far-reaching. Since the leverage points have been selected on a strategic basis, any temporary constraints that arise must be eliminated as a matter of policy. There must be a new organizational policy that reads as follows: **Identify, evaluate and, most likely, eliminate any constraints that have not been selected.**

This policy represents an important decision because it implies confidence in the organizational strategy, and it may require the investment of money to protect this strategy. The benefits should significantly outweigh the costs. As Mark Twain put it in *Pudd'nhead Wilson's Calendar,*

> Behold, the fool saith, "Put not all thine eggs in the one basket" — which is but a manner of saying, "Scatter your money and your attention"; but the wise man saith, "Put all your eggs in the one basket and — WATCH THAT BASKET."

Also keep in mind that this policy requires nonselected constraints to be eliminated. That means you can't assume they don't exist. They will exist, they will have to be eliminated, which means the evaluation process must be going on constantly.

The Market as a Leverage Point

As we've seen, the market should almost always be considered a leverage point. We are in the Throughput World largely because throughput is unlimited. Throughput is unlimited only because the market is (for practical purposes) unlimited; therefore, the market must be a key Throughput World leverage point. Even if you have a resource that has insufficient capacity, there are ways to make more money by being more competitive. Performance relative to the following characteristics can increase or decrease your ability to sell at higher prices, and therefore to make more money:

- Response time
- Quality
- Due-date performance
- Features and options
- Image

In addition, with limited capacity there are also situations where excess capacity can be sold from areas that do not require the constraining resources. In such situations it is perfectly reasonable to compete on price as well as other characteristics. In the example of Figure 17-1, assuming a programming leverage point, projects that did not require the programmer could still be sold. However, as we will see in Chapter 19, their capacity must not be sold completely.

If the market is a leverage point, we want to treat it well. We don't, for example, want late projects. It follows that no internal resource should be committed past its capacity. Jobs that are accepted should not cause overloading of a leverage point. This means that even resource constraints should be subordinated to the market, which means their production should be based on market demand.

Selecting a Strategic Leverage Point

How should you select a strategic leverage point? Many considerations can be given, but they boil down to how the resource will be managed, which boils down to the five steps above.[28] Imagine a particular resource as a leverage point, and decide whether it then makes sense to apply the five steps, making sure to include the market as a leverage point. For example,

- **How would focusing on this resource help you to subordinate to the market leverage point?** Sometimes additional capacity can be gained from resources by focusing on them (see exploit, below), which means you're better able to subordinate to the market. In certain cases it makes sense to locate a strategic leverage point early in the processing chain, in order to minimize work-in-process. Minimizing work-in-process in turn decreases lead times, which increases competitive advantage. In some cases all resources can easily be subordinated to the market for the foreseeable future (meaning they're easily elevated as well). Then it may be appropriate to have the market as the only strategic leverage point.

- **How difficult is the resource to exploit?** If it's simple to exploit, there may be no sense or advantage to the added focus that comes of being a strategic leverage point. Furthermore, it may be difficult to make sure other resources keep up with its increases in capacity. If it's difficult to exploit, if it's difficult to squeeze more out of, the resource could perhaps use the attention that comes with being a strategic constraint.

- **How difficult is it to subordinate to the resource?** It doesn't make sense to have a constraint resource that other resources can't subordinate to. The most obvious case is where the resource has much more capacity than a number of other resources; that is, it's nowhere near a bottleneck. This implies that subordinating those other resources would require purchasing significant extra capacity. As a somewhat more complex example, a resource through which material must flow many times (as happens, for example, in the semiconductor industry) can be difficult to subordinate to, because its schedules are complex and significant inventory must be built for the subordination to occur. This is an area where mistakes are often made, as people may choose to focus on expensive resources (i.e., resources that are difficult to elevate) without noting that they aren't easily subordinated to.

- **How difficult is it to elevate the resource?** If the resource is very expensive and/or it takes a long time to acquire new capacity, it may make sense to call it a strategic resource and treat it with care.

We can derive some general comments from these questions. An external resource (for example, a contractor) will not make a good strategic resource, because you typically have little control over it (you can't exploit it) and you can find others (it's easy to elevate). A resource with lots of extra capacity is usually not a good choice, because it will be difficult to subordinate to. Very expensive or hard-to-get resources may be good choices, because the added attention may gain additional capacity that would otherwise be expensive.

The Five Steps and the Critical Chain

In Part II we discussed the Critical Chain, which is the leverage point for a project. The strategic leverage points for an entire organization are more likely to be the market, plus possibly a very small number of strategic resources. How do these different kinds of leverage points fit together?

Critical Chain scheduling, complete with buffers, enables projects to complete more quickly and more reliably. The competitive advantages it confers are ideal for subordinating more effectively to the market. In Chapter 20 we will look at how scheduling and buffers can be used to combine the market-oriented Critical Chain approach with a focus on overall organizational strategic leverage points, in order to achieve even larger gains than either approach would give alone. Before doing this, in Chapters 18 and 19 we will explore a few more concepts regarding resource constraints.

Key Concepts

- **Select** the leverage point(s).
- **Exploit** the leverage point(s).
- **Subordinate** everything else to the above decisions.
- **Elevate** the leverage point(s).
- Before making any significant changes, **evaluate** whether the leverage point(s) will and should stay the same.
- The market is almost always a leverage point.
- Critical Chain is a technique for subordinating to the market.

Questions for Further Thought

1. Which kind of organization is likely to have more leverage points, a simple one or a complex one?
2. Identify some additional positive effects that might be expected to flow from Figure 17-2.
3. Pick some environments and consider where the strategic constraints should be.

18 Global Improvement: TOC Accounting

I n Chapter 15 we saw examples where traditional cost accounting was insufficient. In particular, we were able to make more money by taking the throughput-oriented approach to Table 15-1, even though we were selling products at below "cost." This is contrary to traditional cost accounting logic.

Assume for a moment that we abandon traditional product costing and pricing methodologies as decision support tools. If they don't work well, this would seem to be sensible — but only if we have something better. What should they be replaced with? What should be the approach to setting product prices and sales targets? We discuss some Theory of Constraints answers in this chapter.

Throughput Pricing

Let's start with a slightly more complicated version of the example from the last chapter.[29] The company diagrammed in Figure 18-1 sells two general types of projects: widgets and gizmos. Data are laid out in the boxes as before. As you can see, widgets are produced in the same way as the projects in Figure 17-1; gizmos require the same resources, in different quantities. The yearly market demand is known precisely, eight of each product. The selling price of widget projects is $50,000 apiece; gizmos are $40,000 apiece. There are 50 working weeks in a year, and yearly operating expenses are $400,000. There is no problem obtaining purchased parts; plenty are available, of good quality, exactly when needed. All the data are available and extremely accurate. There are no disruptions: no sick workers, machine breakdowns, or

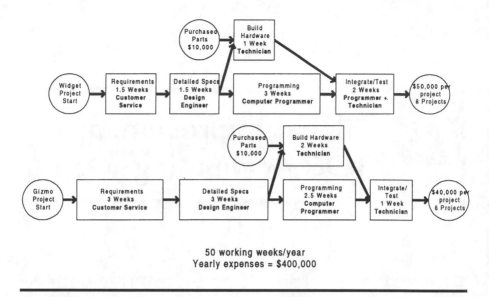

Figure 18-1   Custom Hardware Projects

walkouts. In fact, there's no variability at all. In order that you don't have to schedule each project, we'll also assume that there is already work in the system; projects were being produced at the same rate last year. You have all the data. You are encouraged to treat this as a quiz, and without reading ahead answer the question: what is the maximum profit this business can make per year?

The simplest answer is to calculate the total throughput of the market demands. Remembering that throughput equals sales price minus raw materials, we have

8 widgets × ($50K – $10K) + 8 gizmos × ($40K – $10K) = $560,000.

When we subtract the yearly expenses, we get a nice net profit of $160,000.

However, you might remember that earlier there was an internal resource constraint. That is, there wasn't sufficient capacity to meet all the market demand. We need to check whether that's still the case. To calculate the total resource requirements, multiply each box by the demand on it, and add the results by resource. This gives the load in weeks shown in Table 18-1.

There are only 50 available working weeks in a year. It looks as though the programmer is still constraining the company. That means the company can't make everything it could sell. In order to decide what the maximum

Table 18-1 Resource Requirements (in weeks)

	Customer Service	Design Engineer	Computer Programmer	Technician	Total
Widget	12	12	40	24	88
Gizmo	24	24	20	24	92
Total	36	36	60	48	180

profit could be, we have to decide whether we're making widgets or gizmos. According to any standard measures, the widgets are clearly better. The throughput per project is significantly higher, $40K vs. $30K. The selling price is higher. It takes less total time to produce a widget, meaning the fully allocated product cost is lower. And the overall time to produce a widget is eight weeks, vs. 9.5 for a gizmo. Since the widget is clearly the best product, let's do the calculations. We make 8 widgets, which takes 40 weeks of the programmer's time. The programmer has left over in the year 10 weeks to make gizmos, which means he or she can produce 4. The total throughput is 8 × $40K + 4 × $30K = $440,000. The net profit is $40,000.

Still, we've been claiming that the product costing process is flawed. Suppose we do the calculations with a preference for gizmos. What happens then? The company can sell eight gizmos, and has 30 weeks of programmer time left over. In that time they can also make six widgets. This gives throughput of 8 × $30K + 6 × $40K = $480K, or a net profit of $80K. We've achieved twice the net profit by doing the wrong thing! How could that have happened?

What happened was that we weren't paying attention to the programmer, who was the real leverage point. The programmer's capacity determines the net profit. The question isn't what's the most profitable way to spend the sum of everyone's time; if it were, the product costing approach would be perfect. The question is what's most profitable for the company as a whole. Since the programmer (being a leverage point or constraint) determines overall throughput, the question becomes what is most profitable for the programmer to work on. The programmer can spend 10 weeks to produce four gizmos, or in the same 10 weeks can produce two widgets. That comes out to $12,000 per week of throughput for the programmer to make gizmos, vs. $8,000 per week for the programmer to make widgets. According to this measure, and according to the arithmetic, gizmos are clearly more profitable.

Using throughput dollars per unit of constraint time in this way is called **throughput pricing**. It is a valuable means of comparing the potential throughput of products with the capacity they require. It can be an extremely powerful way of pricing products, especially if the competition is not using it. In cost-oriented companies it is not uncommon to see the throughput per unit of constraint time vary by a factor of 20 or more across different products. For products that go across a constraint or leverage point, maximize dollars per constraint minute. For products that don't go across a leverage point (also called **free products**), by definition you have the capacity to produce them. Any throughput goes straight to the bottom line, as we saw in Chapter 15. If you plan on selling this capacity, chances are one of your resources will eventually become a constraint. If you can identify that resource, it may make sense to use throughput per unit of constraint time as a guide to picking the most profitable products even before the resource runs out of capacity.

Budgeting

One important function of cost accounting is the budget. Budgets are means of controlling and monitoring expenditures. Of course, it is necessary to do this. Problems arise in two cases: when the control of expenditures also acts indirectly and invisibly as a control of throughput, and when the budgetary control works as a lower limit. The first problem can be seen whenever we find an expenditure that we know would produce a big return, and that expenditure is not in our budget. Often, especially in large organizations, either the expenditure is refused or it requires much more justification than seems sensible. The second problem is a frequent result of budgets being based on the prior accounting period's expenses. In such cases the budget becomes a lower limit. It must be spent, because otherwise the next budget will be less. People make sure to spend everything in order to get as much as possible next time.

The answer seems pretty obvious by this time. We can't keep our heads inside the box and focus only on expenses. In order to avoid these budgeting problems, budgets must take into account throughput. The question is how to do it. There's a clue in E. M. Goldratt and J. Cox's *The Goal*: "The actual cost of a bottleneck is the total expense of the system divided by the number of hours the bottleneck produces."[30]

In the above example, the programmer works 50 weeks per year, and the minimum level of weekly throughput (to counterbalance the operating expenses) is $8,000. This means we should be looking for a throughput per constraint week of at least $8,000 in order to meet expenses. In fact, we might add some amount to the $8,000 for targeted net profit to get an actual budgeted throughput per unit of constraint time. If expenses increase as a result of (for example) an elevation decision, we will have to include those expenses in subsequent calculations.

In simple cases this technique can be used alone to set up a budget. Reality is not always so simple. Many things can make budgeting more difficult, including multiple constraints that feed each other (known as "interactive constraints"), but looking at targets for throughput per unit of constraint time can still be useful. If such targets are to be used, they must be used with the revised five-step process with Select instead of Identify. Otherwise old strategies, budgets, and conclusions will become irrelevant as the leverage points change.

Other throughput-oriented budgeting options will depend on the environment. "Throughput justification" procedures should be adopted rather than "cost justification." Additional expenditures can help generate more throughput in a number of ways, including better exploitation of the leverage point, better subordination to the leverage point, better exploitation of the market as a leverage point, and so on. Sometimes a rough estimate of the throughput obtained will have to suffice; this is certainly better than no estimate.

Flexible budgeting, in which variable costs are estimated in order to determine the costs of generating additional throughput, typically do not take constraints into account. They therefore treat as "variable" costs that will not increase with additional product volume. Often workers' salaries are considered variable and management salaries fixed. This is amusing, because frequently there is much higher turnover in management than in labor.

The traditional "cost to benefit" analysis can be performed on expenditures. However, be careful; look at marginal costs, rather than allocated costs, which include things you're paying for anyway. See both the example of Table 15-1 and the example of Figure 18-1 for the traps you can fall into with allocated costs.

In Part II we discussed the Critical Chain as a leverage point. In this chapter we've talked about a constraint resource as a means of focusing the organization. Is one for manufacturing and the other for projects? How can these two different pictures be reconciled? We will deal with this question in the next two chapters.

Key Concepts

- The impact of a sale on productivity capacity must be measured by its impact on the leverage points.
- **Throughput pricing** uses the combination of a product's throughput and its impact on leverage points to determine product prices.
- An effective budget must take into account throughput.

Questions for Further Thought

1. Change Figure 18-1 by increasing yearly demand for gizmos to 16 and adding another programmer. The additional programmer increases operating expenses to $500K/year. What is the net profit now, assuming all other conditions remain the same?
2. Under what circumstances will using dollars per unit of constraint time cause you to take questionable actions?
3. Why is the concept of market segmentation, mentioned in Chapter 15, important if you use throughput pricing?
4. A budget that doesn't take throughput into account can be very precise. A budget that does take throughput into account will almost never be precise. Is this an advantage of the traditional budget?

19 The Project Dice Game

S ingle-project schedules can be very useful. When companies (and/or individuals) have many projects going on at once, single-project schedules may not be entirely realistic. Critical Chain scheduling still works, but the situation becomes more complex when those projects share resources. Let's look at a simple game in order to understand some key concepts.

Playing the Dice Game

To play the Project Dice Game you'll need Figure 19-1, a standard six-sided die, and some pennies (30 should be enough). In this environment pennies represent projects. If a penny is in the "Project Starts" circle, its project hasn't yet been started. If it's in the "Completed Projects" circle, it has been finished. Pennies anywhere else along the processing path are considered to be "work-in-process." The object is to complete as many projects as possible within a 4-week time span.

There are four resources in this environment: A, B, C, and D. A feeds B, which feeds C which feeds D. The work required to complete each project consists of four tasks, each performed by one of the resources. Each task has four job steps, labeled with 1, 2, 3, and Completed. For example, the job steps on task A (performed by resource A) are labeled A1, A2, A3, and Completed A. Work moves through the job steps and tasks following the arrows, from task A through task B, task C, and then task D. There is an unlimited market for projects and no problem with starting new ones. The object of the game is to complete as many projects as possible.

Start with one penny in each of the "Completed A," "Completed B," and "Completed C" circles; put the rest of the pennies in or near the circle labeled "Project Starts." Play goes a day at a time. Within each day you'll roll the die four times, once for each resource. The spots on the die indicate how many

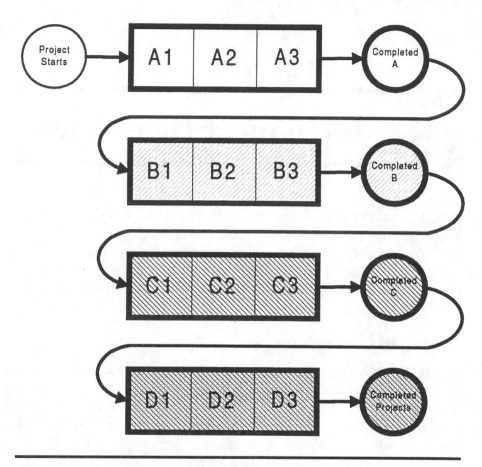

Figure 19-1   Project Game (Balanced)

job steps you can move work ahead for that day, within the resource you're rolling for. To simplify the calculations you should re-roll whenever a one appears. In other words, the acceptable die values are two through six, so there's an average value of four job steps per day per resource. Use of the die simulates the fact that the amount of work people actually complete in a day will vary significantly and unpredictably.

Pennies are moved from step to step within each task according to die rolls. If A initially rolled a three, a penny could be moved from "Project Starts" to the box "A3." There are unlimited pennies in the "Project Starts" circle, so if instead a five were rolled for A, one penny could be moved from "Project Starts" to the "Completed A" circle, and one penny from "Project Starts" to box "A1." B pulls work (if any is available) from the circle "Completed

A" and moves it along to the circle "Completed B." The same kinds of paths are followed for C and D. If a resource has nothing to work on, it cannot "save" job steps; they are lost productivity. Work that gets all the way through D and lands on "Completed Projects" is considered finished and is counted toward your final score.

On any given day a resource may only work on those projects that were available at the start of the day; that's how work is transferred in this company. In order to enforce this, it's simplest to roll the die in the following sequence: day 1, resource D; day 1, C; day 1, B; day 1, A; day 2, D; day 2, C; and so on. In other words, within a day the die is rolled for resources from bottom to top of the picture. That way D (for example) can only work on projects that C had completed before the start of the day. You will have to be vigilant in keeping track of which day it is. On a piece of paper, make a mark each time D rolls, in order to flag a new day.

Before you start the game, estimate the average weekly production you expect from the system, given 5-day weeks. Come up with a prediction of what you'll be able to produce in 20 days (four work weeks). Now run for 20 days. Look at how many projects were completed. Count the number of active projects that exist in the system at the end of day 20. This number is the total WIP inventory. Were there any surprises?

Analyzing the Project Dice Game

The average daily die roll is four (don't forget, we're excluding ones). Since there are four job steps per task, on average we'd expect complete about 1 project per day through each resource, or 20 projects over the course of the game. In reality, you'll average completing just over 4 projects per week, or 16 over the course of the game. Almost never will a player achieve, let alone exceed, the expected 20 projects. Inventory generally triples, although it may increase by anywhere between 50% and 300% or more. What is going on?

In this game we've seen the effects of statistical fluctuations and dependent events on a simple system. The statistical fluctuations cause resources frequently to achieve less and more than the average. When the fluctuations are coupled with dependencies between the events which must take place, at various times some resources are starved and some overproduce. When resources are starved, overall production suffers, because production depends on getting work all the way through the chain. Competitive position suffers, because some resources are overproducing relative to the ones which are underproducing. Therefore, the work-in-process and lead times usually increase significantly. To see the increase in lead times, run the game with a red dot on the first penny processed by resource A, and see how many days it takes for that penny to be processed to completion. Then put a red dot on the first penny released into the system at the start of week 4 (day 16). Continue playing until that penny is completed and see how many days it remains in the system. Since we can sell everything that is produced, this number of days is the lead time for producing projects. How did the lead time change as time went on?

Unbalancing the Capacity

Balanced-capacity systems don't usually perform near their theoretical expectations, for the reasons we've seen. As another experiment, try the same game but with an unbalanced plant, Figure 19-2. We'll decrease by one the number of job steps for the A, B, and D tasks. Our theoretical average for the 4 weeks is still 20 projects; the extra capacity doesn't look terribly useful. What happens when you play out the game?

With the layout in Figure 19-2, playing according to the same rules, it's normal to get much closer to the ideal 20 projects. Sometimes you will even be able to make more. Production has increased significantly.

What else happened? Unless you restricted project starts, significant inventory builds up in front of the C resource. It's very possible that there is more inventory in the system than existed when you played the first game. This makes sense, because over time A and B will consistently bring more into the system than C can push out. C is a bottleneck that restricts your ability to produce more. It is a constraint or leverage point. While we've gained in short-term production, we may have lost the chance to hang onto the market, because lead times are increasing and will continue to increase. This syndrome happens frequently in real life, where elements of a system take on more work than the system as a whole can handle. This can happen with projects, manufacturing, and even children's extracurricular activities.

Protective Capacity and Inventory

What happens if you construct a Critical Chain schedule from the game in Figure 19-1? It's very easy to do, because everything is part of the Critical Chain. There is no place to put feeding buffers without creating space everywhere. This

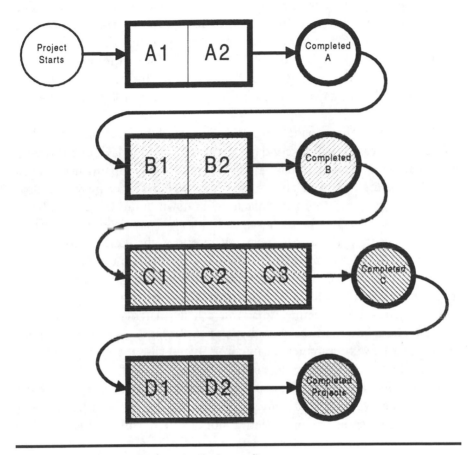

Figure 19-2 Project Game (Unbalanced)

means that everything is a constraining resource, and a disruption anywhere can affect production. We can conclude that in an environment where capacity is balanced, production will not reach the theoretical maximum, and/or inventory will get very high. We saw this when we played the game.

To maximize the throughput of a system like Figure 19-1 or Figure 19-2, it's necessary to keep the most limiting resource (the constraint) busy. In order to do that, other resources must have some extra capacity. The extra capacity needed to keep work available for the constraint resource is called **protective capacity.** This capacity is required in order to distinguish constraints from nonconstraints. There is an inverse relationship between protective capacity and required inventory. On average, the less protective capacity there is available, the greater the amount of inventory that is needed to keep operating at maximum production.[31] Imagine that a resource loaded to 50% of its capacity feeds one loaded to 100% (the constraint or leverage point). If there is an upstream disruption of a day, so that the 50% resource has nothing to work on for a day, it will take a day for that resource to catch up when the extra day of built-up work comes along. Instead of working two days at 50%, the lost day means it must work one day at 100%. It catches up quickly.

Suppose a resource loaded to 90% of its capacity feeds one loaded to 100%. In this case it will take 9 days to catch up with a day's disruption: 9 days at 100%, instead of 10 days at 90%. If there are too many disruptions, the 90% resource won't be able to keep the 100% resource busy, and production will be lost. Protective capacity goes down, more work must be in the system to keep constraints busy, lead times increase. Without sufficient protective capacity, there may be many resources that can be considered constraints. In order to maintain throughput, inventory must go up.

Suppose you are an elementary school teacher, setting up meetings with individual parents to talk about their kids. There are 25 sets of parents, so you have created 25 meeting slots. You call the parents one by one and schedule them into the time slots they prefer. Perhaps you take a few notes on the alternative times parents can come, just in case there are scheduling conflicts. By the time you get to the last few parents, conflicts start to arise. These parents have few choices, and some of them can't make any of the meeting times. Very likely it is impossible to get a reasonable schedule. You might end up spending several meeting slots' worth of time on the phone to find that out. To use our five-step terminology, it is very difficult to subordinate the parents' schedules to the time slots.

There are really two choices. Either get parents with lots of available time (i.e., plenty of spare capacity), or start with many more time slots than there are parents (plenty of spare inventory). Either way, scheduling and rescheduling the meetings would be much easier.

In this example, the parents' schedules are pretty much independent. Now add some more restrictions: certain parents' meetings must occur after other parents. In other words, consider that the parents must work together in chains. You can see that either the available protective capacity or the available inventory must go up even more than before.

Here we can again see the pressure for multitasking. Some spare or "protective" capacity is necessary for most (if not all) resources. But rather than leave idle time so that they're available when really needed, people will be more likely either to make the work last, or to take on other things. They build inventory, in order to keep the capacity in use. And yet for reasons discussed in Chapters 3 and 4 it's much better for people to have some free time than to slow down, or to do frequent switching back and forth between jobs.

This protective capacity could be significant in quantity. It doesn't have to be idle time; it could be spent on non-time-critical research or development, for example. However, it needs to be recognizable as available time. Protective capacity is common with (for example) receptionist jobs, because they aren't highly paid. When a receptionist is a constraint a company looks ridiculous. But in reality most positions need protective capacity. It's important to recognize that hoping for more than a very few people to be busy on scheduled work all the time — really, productively busy — is inevitably going to push lead times out significantly. The apparent gain in productivity is a mirage.

A Resource-Constrained Schedule

Before reading ahead, as an exercise try creating a schedule for the first five projects produced by the game in Figure 19-2. Identify a Critical Chain across all projects at once (even though, as we shall see in the next chapter, this is usually inappropriate). Since everything can be sold immediately, it's most ambitious to assume all the orders are due on the first day of the scheduling horizon. Don't forget to take into account work-in-process, meaning the pennies at the start of the game in the "Completed" circles.

Before buffering, you should get a picture something like Figure 19-3. The tasks of resources C and D are all part of the Critical Chain. C tasks are clearly part of the Critical Chain because C is the most limited resource. After a C task is done we'd like to ship the project as quickly as possible, putting D on the Critical Chain as well. The Critical Chain tasks in Figure 19-3 have bold borders.

Having identified a kind of Critical Chain, we must next decide where the buffers should be put. Project buffers after D tasks would be nice, in order to protect the customers. In this case the customers will take as much as we can give them, as quickly as we can produce it, so the project buffers will mostly be useful for predicting when we can ship. Resource buffers are not needed, because resources C and D do not have any non-Critical Chain tasks.

Feeding buffers are needed where each B task feeds a C task. This is a very important point to understand. If we create these feeding buffers and push the B (and associated A) tasks earlier, we're in effect making sure that

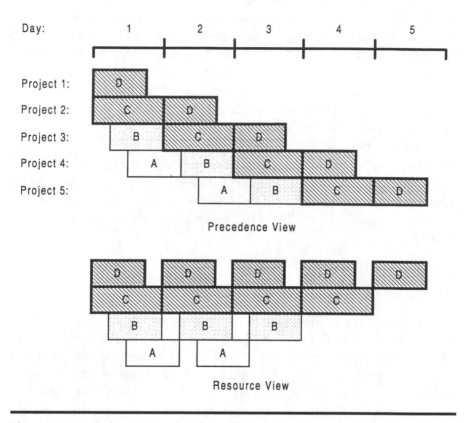

Precedence View

Resource View

Figure 19-3 Project Game Load Leveling

work is available for the C resource. This seems very logical, because we don't have enough of the C resource. We want it to keep busy, because the more it produces the more projects we can complete. On the other hand, beyond what's needed to keep the C resource busy, there's no reason to release additional work. This means we don't want indiscriminately to move the B and A tasks as early as possible. As you might have seen in the last run of the Figure 19-2 game, this can build inventory to arbitrarily high levels. As we saw in Chapter 9, the only reason to build inventory is to protect the ability to produce. At this point you are encouraged to estimate needed buffer sizes and create a full schedule based on Figure 19-3.

We will release material in order to keep C (the constraining resource) working. How much material? On average, A and B can complete one and a third tasks per day. C will on average complete one task every day. However, C (with high rolls) can over short periods outproduce A and B. Without going into any deep calculations, suppose we try to keep 8 days of work (8 days for resource C, that is) between the start of A and the start of C. This should be enough protection most of the time. Since we already start the game with pennies on the Completed A and Completed B circles, we need to allow up to six more projects to exist in between Project Starts and Completed B. Here are some rules that could be used to restrict the flow of work into the systems:

1. In addition to the pennies on the "Completed" circles, start the game with six pennies in the "Project Starts" circle. If during the course of the game this circle is empty, A is not allowed to start any more projects no matter what you roll for it.
2. Each time a penny reaches the "Completed C" circle, put a new penny in the "Project Starts" circle. A is now allowed to start this penny.
3. B, C, and D should work on everything they have.

If you try this, you'll most likely find that C will keep busy, especially after the first week. This means throughput is maximized. It's possible that the luck of the die will for a time cause C to outproduce A and B; this can even deplete inventory to the point where C is idle. There will also likely be long periods where there is plenty for C to work on. You can control the amount of work-in-process by controlling the pennies put on the Project Starts circle. From this you can see the inverse relationship between work-in-process and throughput. More pennies means more work-in-process, but also a greater chance resource C will keep busy, meaning more throughput. Fewer pennies means less work-in-process and shorter lead times, but less chance that C will keep busy and therefore less throughput.

Obviously, real-world projects can be much more complex than Figure 19-2. Nevertheless, this example gives a valuable clue about how to attack scheduling in multiple project environments. Often managers face a conflict between starting a project, in order to make sure it gets completed as quickly as possible, and not starting the project, in order to make sure other projects get completed as quickly as possible. In Figure 19-2 we saw that it's irrelevant to start more work than the leverage point (resource C) can handle; that just builds up more work-in-process. On the other hand, it's also not good to start less work than resource C can handle, because lost time on resource C means lost throughput overall. Resource C determines how things are released, which means project starts should be subordinated to the strategic resource C. Our simple procedure enforces this subordination process.

In Chapter 20 we extend the two concepts of Critical Chain and subordination to the strategic resource in order to deal with more complex multiple-project situations.

Key Concepts

- The more balanced the capacity of a system, the higher inventory must go to maintain productivity.
- Protective capacity is needed on all nonconstraining resources to ensure that they don't become constraints.
- There is an inverse relationship between amount of protective capacity and amount of work in process.
- Projects should not be started at a rate faster than the strategic resource can handle.

Questions for Further Thought

1. You can try other variations of the project dice game. For example, try playing the game of Figure 19-2 with four people, one person at each resource. The people at the resources are measured by how many turns they're idle. The winner at the end of the game is the person who has spent the smallest number of turns with nothing to do. The A resource is allowed to decide how much work is started. What happens to material releases and inventory? Who wins? Who loses?
2. Now allow people to slow down and use any number equal to or smaller than their die roll, if it looks like they're about to run out of work. Is everyone able to stay busy? How is the company production?
3. Appoint a project manager. The project manager will be measured after 20 days by the number of projects completed, minus the number of projects currently in the system. The project manager's job will be to browbeat the workers into not slowing down, and into controlling material releases. However, the workers are not required to show their rolls to the project manager, only to other workers. Now what happens?
4. Experiment by giving the project manager various powers, but leaving the workers' measurements the same. Have a pool of unemployed workers available. Hire a security force.

20 | Multiple Projects

Part II dealt primarily with single-project scheduling using the Critical Chain approach. In many environments this is sufficient. For example, single-project Critical Chain scheduling works quite well when scheduling a move to a new building or the development of a new facility. Single project scheduling can be adequate even in multiple-project environments where the individual projects are, for practical purposes, independent. This can happen, for example, when most of the resources are contracted. On the other hand, frequently companies that spend much effort working on projects have multiple, interdependent projects going on at the same time.

As a management policy there may be resources that are dedicated to specific projects, but there is usually overlap, especially at crunch time. A new project may use existing personnel rather than hiring or subcontracting to get work done. In this situation there can be big advantages to considering resource dependencies between projects.

Planning tasks in isolation means that each task requires significant safety in order to be sure it will finish on time. Planning an entire project together means we can pool the safety into buffers, both reducing overall lead time and increasing predictability. We then have additional benefits of clearer priorities, less confusion, and higher productivity.

Planning projects in isolation means that we are vulnerable to resource contention between projects. The buffers we put in place for the individual projects must be large enough to account for interproject contention for those resources needed in several projects. By planning projects together, we can estimate and evaluate the impact of decisions that might affect several projects. By evaluating the impact of these decisions, events become better coordinated and we need less protection from buffers. In fact, the more globally scheduling is done, the smaller (in total) the buffers need to be, the smaller the lead times, and ultimately the more competitive the company will be.

The last chapters showed the importance of strategic resources and leverage points. Part II showed the importance of the Critical Chain. The strategic resource can determine the overall throughput of the organization; the Critical Chain determines lead times. Relating this back to the five-step process, there are three overriding considerations when planning multiple projects:

1. **The market is a leverage point.** What is the likely impact of a new project on existing commitment dates? What should be the commitment date of a new project? The biggest interproject impact will likely be on the capacity of the strategic resource. We need to subordinate projects to the requirements of the market, and therefore we need Critical Chain scheduling in order to obtain faster and more reliable completion times.

2. **A resource may be considered a strategic leverage point.** If there is such a resource, we need to focus attention on it. It is an organization-wide leverage point. How can we exploit and subordinate to it? This resource dictates the overall throughput of the organization.

3. **A nonstrategic resource can become a constraint.** Where are the problematic resources? We must have the means to identify and ruthlessly eliminate these temporary constraints.

In short, multiple projects often need to be planned together; strategic resources and Critical Chains are both important.

There are many reasons why scheduling more than one project at a time can be difficult:

- The level of synchronization and communication required between project managers, resource managers, and echelons above and below them can be difficult or impossible, especially in large organizations.
- Large amounts of data must be collected and updated in a coordinated fashion.
- There are large technical difficulties in creating schedules.
- The various project and resource managers must have frequent access to the data, not just to browse through plans for their areas of responsibility but to perform updates and make decisions.

This suggests three interrelated elements that make multiple-project scheduling more complex than single-project scheduling: organizational structure, data management, and the technical details of scheduling. Given the opportunities this complexity provides for getting sucked down into the quicksand of detail, there must also be a fourth overriding consideration:

4. **Keep it simple.** What is the purpose of the plan? How much detail is really needed? If people dive into too much detail, if they make things too complex actually to manage, they stop managing and start reacting.

The Strategic Resource Buffer

No scheduling approach that depends only on the Critical Chain would provide the ability to focus on a strategic resource. As we have seen, the strategic resource can determine the throughput of the entire organization. If it is idle, the organization is losing the capability of producing throughput; effectively, the organization is idle. If it keeps busy in a productive way, the organization is producing. This problem is solved by inserting "strategic resource" buffers.[32]

The strategic resource buffer is placed before tasks on the strategic resource in order to make sure the strategic resource has work, and can therefore stay busy. It protects the throughput of the organization. This buffer is created by pushing tasks feeding strategic resource tasks earlier by the duration of the buffer. The strategic resource buffer is much like a feeding buffer, which guards the Critical Chain, except that this buffer ensures that the strategic resource will have work available. The leverage point is able to keep busy.

This buffer is managed in the normal way described in Chapter 9. We check to see how much of the buffer has been used up by late tasks feeding the buffer. If the amount used up is significant relative to the amount of processing time left before the work will get to the strategic resource, there is a problem. If nothing is done to fix the problem, the project will most likely still complete on time, because there is still the project buffer to protect its completion. However, the strategic resource is likely to be idle, which implies that the organization will lose throughput. In other words, future projects will finish later.

Should the buffer be put before a strategic resource task that is on the Critical Chain? This will create space on the Critical Chain, which means the project will take longer. It also ensures that the strategic resource will be kept busy, which means throughput is gained. This is not an easy tradeoff to make. The best solution may be to leave out the buffer (i.e., subordinate the strategic resource to the market), but evaluate the situations as they arise to see if the Critical Chain tasks requiring the strategic resource need more protection.

We have so far assumed one strategic resource. It is definitely possible to have more than one. However, if they are not independent, be careful: as we saw working with Figure 19-1, without people invoking Parkinson's Law it isn't feasible to have several resources working all the time on important tasks. Furthermore, the more places you must put your attention, the less focused you are.

The strategic resource buffers must be monitored by the project manager, because he or she is responsible for making sure that the work arrives at the strategic resource on time. However, there should also be a resource manager, who is responsible for managing the strategic resource directly. This person will need to collect and evaluate strategic resource buffer information from all project managers.

With the strategic resource buffer in hand to watch over the strategic resource, we're now in a position to move on to some different scheduling approaches for multiple projects.

Approach 1: All Together

This is the most obvious approach to scheduling multiple projects. Take all the tasks for all the projects and schedule them at the same time. We tried to do this in Figure 19-3. In Figure 19-3 we didn't need strategic resource buffers because the feeding buffers served that purpose. We can't always count on the feeding buffers being present to protect the strategic resource, if the strategic resource tasks aren't on the Critical Chain.

Scheduling

Level the load (resolve resource contention by shuffling tasks) and place tasks at their latest possible start times, taking into account the required project completion date. Flag all tasks with no slack to the past or future; these are the Critical Chain tasks. Be sure to put strategic resource buffers before strategic resource tasks on the non-Critical Chain. Put in resource and feeding buffers to protect the Critical Chain, and project buffers after the different projects.

Unfortunately, reality is seldom as simple as Figure 19-3. For projects not due for some time you may have no Critical Chain, because all tasks will have slack to the past or future. For example, if you have no work next year and a project to work on in 2 years, all tasks of that project will have slack to the past. Without a Critical Chain, there will be no feeding or resource buffers. Without feeding or resource buffers, the project buffer will have to be significantly larger.

Managing Buffers

The project manager needs to keep track of all four types of buffers: project, feeding, strategic resource, and resource. The manager of the strategic resource must keep track of the status of the strategic resource buffer, in order to try and ensure that the resource keeps busy. With this approach there may be entire projects having no Critical Chain tasks. If that's the case, simply having a project and strategic resource buffers may not be good enough for management purposes; this is a real problem with the "all together" approach.

Adding New Projects

This gets tricky. Adding tasks for a new project, into whatever available time slots they'll go, could change the entire schedule, including the Critical Chain and the buffers. Technically it's probably feasible if done with a computer. However, Critical Chain schedules are supposed to provide stability, so these kinds of schedule changes run against that philosophy.

Conclusions

In a complicated environment, this approach will not work. Organizationally it causes many interactions between projects and resources, meaning that management according to any traditional structures would be virtually impossible. A large amount of data must be maintained in a synchronized fashion. The Critical Chain will have a tendency to change when new tasks are added that fill available times of resources.

In a simple environment, such as Figure 19-3 (a real-life example might be a tool or die-making projects), this approach could actually be very good. The Critical Chain will consist primarily of strategic resource tasks, plus tasks fed by the strategic resource. Such simple environments might not have resource and project managers, meaning some of the organizational obstacles aren't present. Meanwhile, the focus on the strategic resource and fast-as-possible project completion is maintained. The four "overriding considerations" discussed above are taken into account. In such simple environments the "all together" approach is very similar to the manufacturing scheduling approach "drum-buffer-rope."[33]

Approach 2: Successive Projects

This approach calls for scheduling projects individually, one at a time, including identifying a Critical Chain and adding all four kinds of buffers. Then the project schedules are put together successively into a composite schedule, with resource contention between projects resolved a project at a time. Buffers are updated and managed for individual projects.

Scheduling

Schedule a new project by itself, according to the approach in Chapters 11 and 12. When identifying the Critical Chain, you will be finding the longest chain through the project. Even if the project were due in a hundred years, it would still have a Critical Chain. When identifying buffer points and inserting buffers, be sure to include strategic resource buffers.

For interproject scheduling, you will be maintaining a composite schedule, which has the expected start and finish times of all tasks based on the current status of all tasks. It's important that this interproject schedule have the currently expected start and finish times, not the originally scheduled times, because that's a more realistic picture. Add the new project to the end (the latest point) of the composite schedule and push the project as early as it can go, but no earlier than its desired completion date. Individual tasks on the new project can skip over tasks on old projects, if the changes won't cause problems with the old projects, and if the change will help make the new project closer to its completion date. Tasks on old projects can be pushed earlier, if the slack is available and if the result doesn't cause a big disruption in those projects.

The new project schedule can then be pulled back out of the composite schedule for the project manager to use in managing the individual project.

Managing Buffers

The project manager must keep track of the four kinds of buffers, as already discussed. The manager of the strategic resource must always keep track of the current status of the strategic resource and its buffers.

Adding New Projects

This is taken care of under "Scheduling." A new project is fitted onto the end of the existing composite schedule. Someone must manage the composite schedule and make sure that tasks in it contain expected start and finish times.

Conclusions

This approach is much better than the previous one. Organizationally it's much simpler to deal with; project managers are responsible for individual projects that have individual Critical Chains. Resource contention between projects is minimized. Project schedules will not be changing frequently. Normally schedules will only change when it's desirable to move a previously scheduled task earlier, in order to make room for a new task on a new project.

There are also a potential drawbacks to this approach. First, it requires that all the up-to-date project data exist in the same place. That project data can't be just the original schedules, because tasks in ongoing projects will be late or early, affecting resource contention. If tasks in Project A are late, that should be taken into account when scheduling Project Z. Of course, projects A through Y don't actually need to be rescheduled, because they're protected by buffers.

There is another related drawback: the emphasis on large amounts of data might lead one actually to believe all the data, or to attempt to make it all believable. This is a trap that should be avoided, as per overriding consideration #4: keep it simple. In most situations the inherent imprecision of the data makes it a waste of time to resolve resource contention everywhere.

Approach 3: The Strategic Resource

The strategic resource approach is very similar to the "successive projects" approach, except that resource contention is not resolved between projects for all resources. Contention is only resolved between projects for the strategic resource. Resource contention within projects is resolved as usual, across all resources. We make the assumption that either the only significant resource contention between projects will be contention for the strategic resource, or that by resolving contention on the strategic resource, we will also create more space in which other resource contention can be resolved. That means we can schedule a project in isolation, as long as we adjust our project's schedule based on the availability of the strategic resource.

Scheduling

First, there must be a resource manager in charge of keeping an up-to-date strategic resource schedule. This is the schedule against which new projects will be matched, in order to resolve interproject contention for the strategic resource. This schedule should have expected start and finish times of tasks on the strategic resource. It should be the most current picture.

Schedule a new project by itself, according to the approach in Chapters 11 and 12. The Critical Chain is the longest chain through the project. Be sure to place strategic resource buffers. Now fit this project into the strategic resource schedule, and push the project as early or late as necessary. Strategic resource tasks on the new project can skip forward or backward over strategic resource tasks on old projects, if the changes won't cause problems with the old projects, and if the changes will help the new project. Tasks on old projects could be pushed earlier, if the slack is available and if the result doesn't cause a big disruption in those projects.

The manager of the strategic resource must then be given the strategic resource task times from the new project, in order to track the strategic resource's progress.

Managing Buffers

Buffers are managed as with the other approaches.

Adding New Projects

This is taken care of under "Scheduling." A new project is fitted onto the end of the existing strategic resource schedule.

Conclusions

This approach has a number of advantages over the others. It is relatively simple. It gives a lot of focus to the strategic resource. It connects project managers through the most critical resource, while not requiring a lot of centralized data.

When starting out with this approach, it may be necessary to schedule many projects at once, rather than adding single projects one-by-one. In this case load must be leveled on the strategic resource for many projects at once, taking into account various requirements of the different projects. In this case a computerized system may be necessary. In any event, simplicity is even more important.

This approach also has a big disadvantage: there can be interproject resource contention for other resources. There are two parts to this disadvantage. First, psychologically it can be difficult to convince people that resource contention should be resolved on only one resource, despite the fact that in reality contention often isn't resolved anywhere.

The second drawback is more concrete. There can be real contention with other projects for nonstrategic resources, especially for resources that may be used both early and late in projects. Fortunately, there are a number of ways this can be dealt with. First, of course, buffers will need to be somewhat larger. Second, we recall the rule from Chapter 17: **Identify, evaluate, and, most likely, eliminate any constraints which have not been selected.** In other words, the act of selecting a strategic resource must be coupled with management commitment to minimize contention on other resources. And third, it is possible to move toward the "Successive Projects" approach by scheduling additional resources between projects. If a resource is really problematic, it can be scheduled together with the strategic resource, although without its own strategic resource buffers. Interproject resource contention can then be resolved across all scheduled resources. In other words, the "Strategic Resource" and "Successive Projects" approaches are the end points of a spectrum, and it's not necessary to have an approach at one end of the spectrum or the other.

Recommendations

The "All Together" approach will be too complicated to be used unless the projects are very simple. The "Successive Projects" approach can work, and in fact is used in environments as diverse as aircraft repair and research and development. However, the requirements of the centralized database may make it difficult or impossible, especially when first implementing TOC, and the data probably aren't good enough to use to resolve inter-project contention for many resources. The "Strategic Resource" schedule is probably the best way to start. Additional resources can be scheduled but only if it's necessary to minimize resource contention. Resolving this contention will also allow buffers to be shortened, as they don't need to protect from as much of the "uncertainty" caused by predictable contention.

Key Concepts

- The market is a leverage point.
- A resource may be considered a strategic leverage point.
- A nonstrategic resource can become a constraint; this should not be allowed to happen.
- The strategic resource governs the throughput of the organization.
- Critical Chain scheduling is needed to minimize lead times and protect due-date performance.
- Strategic resource buffers are needed to protect the strategic resources, and therefore overall organizational throughput.
- The "strategic resource" approach to multiple project scheduling, in which interproject resource contention is resolved on only the strategic resources, can be a feasible starting point for multiple-project scheduling.
- Significant interproject resource contention might require moving toward the "successive projects" scheduling approach.
- Keep it simple.

Questions for Further Thought

1. Think of some possible problems and pitfalls if a single project manager in a multiple-project environment starts scheduling according to the Critical Chain concepts.
2. Is it relevant to select a strategic resource if all of an organization's resources have plenty of capacity? Why or why not?

Endnotes

1. See, for example, 40–41, Goldratt, E. M. and Cox, J., *The Goal*, 2nd ed., North River Press, Croton-on-Hudson, NY, 1992; Goldratt, E. M., *The Haystack Syndrome*, North River Press, Croton on Hudson, NY, 1990, 10–12.
2. These fundamental measurements are defined slightly differently in Goldratt, E.M. and Cox, J., *The Goal*, 60–62, and have been discussed many other places since.
3. We can also call an improvement something that increases the value of the organization. This will usually happen through an increase in generation of throughput, an increase in the ability to generate throughput, or more efficiency in the generation of throughput; meaning that most cases are taken care of by the three basic measurements.
4. For a short, popularized discussion, see Gina Kolata, "Ethicists Struggle to Judge the 'Value' of Life," *The New York Times*, November 24, 1992, C3.
5. See Goldratt, E.M., *The Haystack Syndrome*, 32–33.
6. For another discussion of topics dealt with in this chapter, see Goldratt, E. M., *The Haystack Syndrome*, 47–51.
7. This example is based on one developed by Robert Fox and presented in *Machining Today* magazine, Special Issue, Johnson Hill Press, Fort Atkinson, WI, 1996, 10–11.
8. The TOC meaning of "market segmentation" is an ability to sell the same products at different prices; really, a kind of product segmentation. Airlines do this all the time. This is different from the standard marketing definition, which is the breaking of a consumer market into smaller, more homogeneous groups. Dr. Goldratt has proposed three "rules" for market segmentation:
    - Segment your markets, not your resources. In other words, give yourself as much flexibility as possible in both how you sell and how you produce.
    - Segment your markets in such a way that not all segments will go down at the same time. This is a means of protecting against market downturns.
    - Leave something on the table. That is, don't take all possible business just because it's there. A monopoly is not good, because it carries significant obligations that cut down on your flexibility.

9. These graphs are intended to be indicative rather than definitive. A thorough analysis would require examining additional factors, such as profitability figures of U.S. and Japanese auto companies. (*Source:* AAMA, *World Motor Vehicle Data,* American Automobile Manufacturers Association, Detroit, 1996.)

10. AAMA, *World Motor Vehicle Data,* American Automobile Manufacturers Association, Detroit, 1986, 1989, 1992, 1996.

11. Yearly data for the U.S. are only available starting in the mid-1970s, so this graph gives biannual data points only. (*Source:* AAMA, *World Motor Vehicle Data,* American Automobile Manufacturers Association, Detroit, 1986, 1989, 1992, 1996.)

12. A more complete discussion of the Throughput World can be found in *The Haystack Syndrome,* 52–58 and following.

13. Goldratt, E.M., *The Haystack Syndrome,* pp. 52–57.

14. Peter Senge writes about "The Principle of Leverage" in Chapter 7 of *The Fifth Discipline,* Doubleday/Currency, New York, 1990. The concept is very similar. The Throughput World mindset assumes there **must** be major leverage points, and looks for them.

15. See Murray Gell-Mann, *The Quark and the Jaguar,* W. H. Freeman, New York, 1994, 92–100, for an interesting informal discussion of this topic.

16. Goldratt, E. M. and Fox, R. E., *The Race,* North River Press, Croton-on-Hudson, NY, 1986, 179.

17. *Ibid.*

18. *Ibid.*

19. It was not, however, a viable option to send nothing at all.

20. This was related in a paper by General Patricia Hinneburg, Joseph Black, and Captain William Lynch entitled "Lean Logistics" presented at the 1996 APICS Constraint Management SIG meeting.

21. Hindle, D., "Measuring the Utility of Health Care," University of New South Wales, Australia, 1996.

22. Michael Rothschild, *Bionomics,* Henry Holt, New York, 1990, 192–193.

23. Sun-Tzu, *The Art of War,* translated by Samuel B. Griffith, Oxford University Press, Oxford, 1963, 98.

24. In the market, because the demand of 18 projects per year can now be more than satisfied by existing capacity.

25. For a discussion of the original five steps, see Chapter 11 of Goldratt, E. M., *The Haystack Syndrome.* The five-step process has appeared frequently in subsequent literature; see, for example, *Re-Engineering the Manufacturing System.*

26. Few things are more pleasant for a manager than having employees who don't need to have every detail specified; that is, employees who can be subordinate to the manager's objectives rather than to the manager herself.

27. The "miracles" from Chapter 7 could also be called "injections."

28. A detailed discussion of the considerations involved in selecting a strategic resource is beyond the scope of this book. For example, V, A, T analysis can give important clues to where control points should be located; see Umble, M. and Srikanth, M., *Synchronous Manufacturing,* Southwestern Publishing, Cincinnati, OH, 1990. In most situations this kind of detail isn't necessary.

29. The form (not content) of this example is based on one presented in Goldratt, E. M., *The Haystack Syndrome*, 66–78.
30. *The Goal*, 2nd ed., North River Press, Croton-on-Hudson, NY, 1992, 159.
31. Also see 320–326 of Goldratt, E. M. and Cox, J., *The Goal* for a discussion of the relationships between lead times and "spare" or protective capacity.
32. This is also referred to as a "bottleneck" buffer. See Goldratt, E. M., *Critical Chain*, North River Press, Great Barrington, MA, 1997, 236.
33. This approach is described in detail in Goldratt, E., *The Haystack Syndrome*.

IMPLEMENTATION IV

ISSUES

We have been proceeding through the miracles indicated in Chapter 7 as being necessary. In Part II we presented a new approach to project scheduling and logistics, the Critical Chain approach, which means

1. **We have an approach to scheduling and logistics that protects us from the effects of Murphy's Law.**

   Certainly it's important to be able to deal with uncertainty. But more than that, we needed this as a kind of demonstration that new techniques can be invented that really enable improvement. In a way, this miracle is a demonstration that the other miracles might also be possible. We have created means of achieving dramatic improvements — improved abilities to focus people's attention by avoiding multitasking and identifying a Critical Chain, reduced lead times by pooling safety into buffers, and so on. We have a mechanism, the buffers, that tells us what areas we need to focus on for global improvements, and that allow us to communicate this information.

   Our second miracle requires understanding the global viewpoint:

2. **People are focused on global (system-wide) improvements rather than local ones.**

In Part III we defined what global improvements are, and looked at a number of techniques associated with global improvement. Probably the three most important concepts are the goal, the meaning of throughput, and the five-step focusing process.

The five-step focusing process can be described in "the terminology of the improvement process itself"[1] as three questions: what to change, to what to change, and how to cause the change.

Part I of this book deals with the question "What to change."

Parts II and III deal with the question "To what to change."

"Thoughts are but dreams till their effects be tried..."[2] We haven't done anything useful if we can't implement the concepts. Miracle 3 is about how to cause the change:

3. **Everyone understands and accepts the policies, procedures, and measurements that apply to them.**

It's not feasible in one book to define specifically to what to change for everyone, because everyone is different. We must still think through the needed changes to policies, procedures, and measurements. That means it's even more difficult to talk specifically about how to implement the needed changes, and how to cause people to work together effectively afterward. Therefore, we won't be able to give detailed answers to the third question "how to cause the change." We can give a number of ideas and guidelines that should be useful during the implementation process. That is the purpose of Part IV.

21 | What Is a Schedule?

It's not enough to plan; one must also act. In this chapter we discuss what it means to release schedules for people to follow, and the attitudes needed to make Critical Chain schedules work. This means we are talking about operational schedules, not rough schedules used to prepare bids.

In the example of Figure 19-2, we created some rules that dramatically simplified the scheduling approach. Scheduling was reduced to controlling the flow of work into the system, by controlling what went into resource C. We went to all the trouble of identifying the Critical Chain in Figure 19-3. We produced a detailed schedule, and put in buffers after that. And yet we didn't even need to use the schedule. What's going on? What is really needed from the schedule? If we can boil a schedule down to a few simple rules, it's probably time to talk about what information is really needed from a schedule, and what we should do with it.

The most important question is "What is the schedule designed to accomplish?" We have the answer to that at the start of Chapter 10. We want to **maximize throughput,** meaning get as many projects completed as quickly as possible. We want to **minimize inventory,** meaning keeping the smallest possible amount of work-in-process, and hence the shortest possible lead times.

Usually there are many different levels in an organization. The information needs of these different levels are not the same. These needs should be thought through carefully, in terms of both what is and is not needed, and the likely effects of providing different kinds of information. Here we present some ideas for three levels in the organization: the workers, the project managers, and the resource (or line, or functional) managers.

The Worker's Point of View

Let's start with the point of view of the worker. A number of schedule components are clearly needed:

- **Start times for tasks with no predecessors.** Tasks without predecessors are sometimes called "gating" tasks. They function as gates, to control how much work goes into the system. They are equivalent to resource A in Figures 19-1 and 19-2, and they function like the faucet in Figure 3-1. It's very important to control the flow of work, so that too much isn't released. Otherwise lead times rise, confusion increases, and competitive advantage goes down.

- **The priority, if multiple tasks exist.** We try to avoid multitasking by resolving resource contention in advance. Sometimes this is not really possible or even desirable. When people do have more than one job to do, the priority should be clear. Then some of the worst problems with multitasking, such as setups and worst-case lead times, are minimized. Generally speaking, Critical Chain tasks will have highest priority, and then the priority will be according to which task is needed first to avoid problems in the buffers. As schedules are executed, the buffers can be used to determine priorities: the tasks causing problems in the project buffer will be highest priority, then tasks causing problems in the feeding buffers.

- **Who gets the work next.** Once a person finishes a task, what is supposed to happen with it?

- **Sometimes, approximately when your next job is coming and what it is.** It can be important to know when work is coming, so that preparatory steps can be completed. This can be very important for purchased equipment and contracted tasks, so that the vendor can schedule the time for the work to be done.

- **Sometimes, how urgent a job is.** Occasionally, if a task or path is causing a buffer to be used up, people will need to put forth valiant efforts to get that work completed quickly.

- **Task descriptions and requirements.** Strictly speaking this is not part of the schedule, but since it may be part of the prerequisite tree we include it anyway.

There some things that are not needed and that, typically, workers think they do need to see:

- **Start and finish times for tasks.** The start times can work to prevent work from starting early. They indicate when management expects the work to start. We want to allow early starts for everything except the gating operations. Finish times are aids to multitasking — they indicate how long you can postpone completion of a particular task. They can function as self-fulfilling prophecies. Start times can inhibit starting early.
- **Task durations.** These also function as self-fulfilling prophecies.

The idea of no start and finish times runs against the grain for most people, and requires some further explanation. Suppose that in Janet's first schedule from Chapter 8 (see Figure 8-2) she had completed her first errand in 5 minutes. If the start of the second task had to be 10 minutes after the start of the first, she might then feel compelled to stop for coffee. If the coffee shop is slow, this could actually make her late. And later, when the inevitable negative fluctuations occur, there is nothing to counterbalance them.

What about a finish time? Yes, that gives people an idea of what is expected and what they have to fit their work into. Is this good for the project as a whole? The real "benefit" of a start plus finish time for someone's actions is that they can figure out whether there's time that can be spent on something else. They might spend "extra" time on procrastination or perhaps attending lower-priority meetings. Available time is necessary for making scheduling decisions, but is a terrible basis for prioritization decisions. The real global impact of using up extra time is rarely checked.

The next obvious question to be answered is whether to withhold these data. Just refusing to give workers start times (except for gating tasks), finish times, and task durations is a good way to get a rebellion. The best approach is through education: people must understand the irrelevance of these data. That way they won't feel a need to have them, and it won't matter whether they see the times or not. Then the schedules don't have to be locked up at night, and everyone can be more comfortable.

Resource Manager Information Needs

Resource managers are responsible for the technical aspects of accomplishing specific tasks — how the task will be done and with what resources, taking into account any limitations the project may have. Resource managers need to monitor the status of tasks in order to know where they will have spare capacity (available resources). The resource managers consequently need to know

- **When tasks are late or early, and by how much.** Probably this will be needed in order to evaluate worker capabilities, so that they will be assigned to appropriate tasks. It is especially important to make sure gating tasks do not start too early.
- **How important is any lateness or earliness, that is, the status of the buffers being fed by the tasks.** Significantly late tasks may require more resources.
- **The currently expected start and finish times of various tasks.** This allows resource managers to track and predict resource availability.
- **The status of resource buffers for their resources.** Even if the resource buffers are only wake-up calls, this will give them notice that resources must be prepared to start critical tasks.

Project Manager Information Needs

The project managers must make sure their projects stay on track. They are directly responsible for the budget, the schedule, and ultimately the performance of the completed project. They need to know what's going on in order to take actions to make sure they meet their commitments. There are many data points that project managers may need in order to properly manage projects. They need to know at least the following:

- **Project buffer status.** How much of the project remains to be completed, and how much project buffer remains? Based on this, is there a serious problem? What tasks, if any, are holding things up?
- **Feeding buffer status.** Are there feeding buffers that are in trouble? What is the likely impact on the project buffer?
- **Critical Chain tasks and resource buffer status.** When will resources be needed for Critical Chain tasks? When will the resources be available?

There are also a few rules that project managers should pay attention to:

- **Don't reschedule frequently!** Only if the project is in real trouble, meaning the project buffer is in real trouble, will it make sense to reschedule. Remember, rescheduling can change the Critical Chain, and thereby change the focus of the entire project team.
- **Don't worry about a late task unless it is important.** The buffers should tell you what is important and what is not.

- **Evaluate decisions not just on costs, but also on eventual performance of the project.** If, for example, it makes sense to use idle resources to hurry the project along, why not? What are the real tradeoffs, from a Throughput World point of view?
- **Keep it simple!** As usual.

Critical Chain Principles

The above guidelines for who needs what information are important to study, but they are not comprehensive. They don't cover all situations or all people. It's useful also to discuss Critical Chain schedules in terms of principles that can be more easily accepted and generally applied.

1. *Do not communicate misinformation.* Misinformation is not just communicated when we say something that isn't so. It is also communicated when we pretend to know something that we don't. This is a fundamental and very difficult shift in the way we look at schedules. The principle is that **we don't assume any knowledge we don't have.** If schedules are imprecise, we can't assume precision. Pretending we have precise information means either we ignore it, which is difficult, or we rely on it, which can be disastrous. In either case we're not better off. A logical consequence is that people need to understand both what is precise and what is not in the tasks and information they're given.

   Sometimes people insist on being given misinformation. For example, top management might ask for due dates, but object to buffers as "planning to fail." Why not give them unbuffered completion times (5% chance of finishing by this time) and buffered completion times (95% chance of finishing by this time)?

2. *Communicate task information in a global context.* Very often there's a management tendency to give people instructions that don't fit into a global context. That's much easier than having to explain the global context, assuming the managers understand it in the first place. It's sometimes easier to think of people as machines or computers who will follow instructions. Even with unambiguous computer languages, it's impossible to instruct computers without constructing bugs, because they can't know what we want. It's even more difficult when treating people as machines. There's a well-known impossible

task beloved of technical people: write a procedure for someone to construct a peanut butter sandwich. Of course, any instructions given can either be questioned for clarity (what does "spread" mean?) or precision ("what kind of bread?"). There is no end to the required level of detail if the listener doesn't understand, or refuses to accept, any minimum set of assumptions.

People need to understand what is "good enough." They need to understand how what they're doing fits in. They need to be "empowered" sufficiently so that they can subordinate themselves to the global objectives (see also Chapter 17). To do that, they don't have to know everything about everything. They do need to know how their tasks fit in. The prerequisite tree can communicate some of this; a broader schedule with precedence and resource views can also be useful. Making sure that everyone understands all the relevant customer requirements is very important.

3. *When you have scheduled work, do it as quickly as possible.* When we give someone a schedule for their work, what do we want them to do? Ideally, they would work as quickly as possible. This means projects complete more quickly and WIP is minimized.

What happens when there is no scheduled work? Several examples in Chapters 17 to 20 would lead us to expect that to happen. If people always have urgent tasks, they will always be late. If we want them not to be late, sometimes they cannot have scheduled tasks. When there are no scheduled tasks, unscheduled tasks can be performed, such as training, personal projects, or tidying up. Management needs to know about unscheduled time; not to identify people to lay off, but in order to be able to use that time when it can contribute to throughput. We also need to keep in mind the admonition from Chapter 3: when there is more than one task that someone can work on, the priorities need to be very clear. Otherwise the projects associated with all the tasks may suffer.

4. *When you don't have work, don't pretend you do.* Don't create work when there is none. Don't obscure the question of how much capacity you have. These last two rules are sometimes called the **roadrunner** mentality, after the cartoon character. The cartoon roadrunner has two speeds: full speed ahead, and dead stop. The implications of this are discussed more below.

5. *Start as early as possible, if (and only if) it helps the project or the organization.* The key in reducing system-wide work-in-process is to

control the flow of work into the system. That means the operations that control this flow, the gating tasks, must have start times. Otherwise they won't know when to introduce the work.

On the other hand, we want the possibility of early starts, especially on the Critical Chain. If you're waiting for someone else to finish before you can start your work, your start time should be governed by when they finish. It's not possible for you to have a precise start time unless their finish time is precise, which isn't likely. The general principle is: **gating tasks should not start before the scheduled start time; non-gating tasks should be started as soon as they can when work becomes available.** Once again, the roadrunner mentality.

6. *Give people the information and understanding they need.* Many of the kinds of needed information have already been discussed; other requirements may arise based on the situation. In addition to information, people must understand the meaning of the information. They need to know what schedule information means, and what it does not mean. This implies not just education, but discussions and review. Everyone can contribute to the development of common sense.

The Pit Crew

In an automobile race, the pit crew services the car — changing tires, adding fuel, possibly even changing an engine. They work intensely for just a few minutes the during the race. While they may have other duties, their main purpose during the race is to make sure the car spends as little time as possible in the pit. They must be ready to act quickly when the car enters the pit, which means they must be ready in advance. Every second saved in the pit is a second saved in the race.

The tasks that are part of the Critical Chain are much like the race track; the status of the tasks being worked on is the position of the race car. By definition, time saved on a Critical Chain task is time saved on the project. Non-Critical Chain tasks must be ready and waiting when the Critical Chain tasks need them. Everyone must work together for the project to continue at top speed. Some of this preparation is done automatically by the resource and feeding buffers. But while the race will be run on the Critical Chain, everyone can have an effect on the outcome. Everyone must understand how their work contributes to winning the race.

A number of things must go along with this pit crew mentality to make it work. First, everyone — especially management — must maintain a global perspective. The customer is the racing official, and the race isn't over until the official says it's over. This in turn implies that everyone has sufficient globally oriented information, as discussed above.

Second, everyone must understand and maintain appropriate intensity. If pit crew members lose intensity while the car is out of the pit, we don't care. Losing intensity while it's in the pit can be costly. One difference between projects and car races is that races don't take years to run. Burnout can easily happen when everyone is asked to make sacrifices all the time, especially since often the sacrifices prove unnecessary.

Third, a Throughput World mentality must be maintained. It's easy to leave the pit crew and slip into a Cost World mentality. Suppose, for example, there's a key worker who has a Critical Chain task. On examination, it appears that 30% of her time is spent doing clerical work that could be done just as well by someone else. If someone else could do that work, the key worker could finish the task in seven-tenths the time. Suppose even that this clerical work requires the additional person to be available all the time at a moment's notice, to the extent that he or she would be doing productive work less than half the time. Traditionally this would be called a waste. Usually it wouldn't even be considered. Applying the pit crew mentality, the tradeoff would be expressed as the operating expense of an extra person vs. decreasing the overall project time by 30% of the task's time. The possibility becomes much more interesting.

Key Concepts

- Workers need to know what to work on, in what priority, and perhaps at what intensity.
- Workers do not normally need task start and finish times or durations.
- Resource managers need to know the status of the resource buffers and the resources.
- Project managers need to know the status of all buffers.
- Project managers should avoid frequent rescheduling.
- Everyone should have the roadrunner mentality.

Questions for Further Thought

1. Match the needed schedule components given in the section "The Worker's Point of View" with the rules used to control Figure 19-2.
2. What understanding will top management need in order for project managers to be able to follow the principles in the section "Critical Chain Principles"?
3. Think of some other analogies for the Critical Chain besides the race track.

22 | Measurements

Earlier we discussed the primary globally oriented measurements: throughput, investment, and operating expense. A Throughput World implementation is not possible without some access to these measurements, however informal or imprecise. We also talked about related global measurements such as work-in-process and due-date performance. These are very closely tied to the market, and hence to throughput.

Local measurements, which measure parts of an organization at a non-global level, are also extremely important. In the current reality tree of Chapter 5 we saw some problems arise from locally measuring how busy people are, and whether they meet their individual commitment dates. That alone should make it clear that local performance measurements are an important part of any implementation.

A complete discussion of organizational measurements is far beyond the scope of this book; fortunately, there are a number of useful references for TOC measurements.[3] There are also some basic concepts that should be reviewed here, and some important applications of measurements to the implementation of TOC in project environments.

Measurement Basics

Measurements are not just a barometer of how the organization or people in it are performing. Measurements — formal and informal — are the means people have for knowing what to do and when to do it. Informal measurements are sometimes more important than formal ones, because everyone is constantly measured and measuring. For example, children are constantly measured, and different measures have different rewards or penalties. Many of the measures are not just formal, but unintentional. If children are given attention (a reward) for acting out, they may continue to act out. If children are punished for believing in themselves or their own feelings, they will learn poor self-esteem. If workers are given a frown if they aren't doing anything, they will find work to keep busy. "Tell me how you measure me and I'll tell you how I'll behave."[4] Existing measurements, whether formal or informal, must be put on the table so they can be understood and evaluated.

Because there are many uses for measurements, there can be many types of measures. In fact, everyone who has a stake in a company — customers, vendors, shareholders, managers, programmers, and so on — might be measured differently by the company according to the organizational objectives. Customers might be measured on how much they buy, how quickly they pay, and how little they complain. Looking in the other direction, members of those groups also have different ways of measuring the company. Customers will measure the company based on their satisfaction with the product in terms such as price, quality, delivery, and performance. It is important to understand where a given measurement is being applied.

The effects of any measurements should be evaluated to make sure that the desired results are achieved. This is very important. Measurements are a cause. The effects should be evaluated, preferably in advance. Look back at Figures 5-2 to 5-4 for a moment. What is the effect of evaluating individuals, or allowing individuals to evaluate themselves, using measurements that don't have a direct global connection? Their actions will not have a direct global connection.

To summarize, for any measurement you must understand its

- Derivation: specifically, what goes into the measurement;
- Application: how the measurement will be applied; and
- Effects: what global effects you expect to achieve with the measurement.

It's a useful exercise to see if you understand these points with your current organizational measurements.

Measurements must also be communicated. There is a second part to the warning given above: "Tell me how you measure me, and I will tell you how I will behave. If you measure me in an illogical way ... do not complain about illogical behavior."[5]

Global Measurements

Any local measurements must have direct cause-and-effect relationships with the global measurements. Consequently, the first order of business is to establish the global measurements. The global measurements from the perspective of the owner(s) of the company are throughput, operating expense, and investment, as discussed in Chapter 14. Besides these financial measurements, there are other useful global measurements. For example, due-date performance and work-in-process are useful measures of how well the organization is subordinating to the market.

While the principles behind the global measurements can be explained, it's very difficult to get a computer programmer or design engineer or lathe operator to think in terms of them, for several reasons. First, workers don't normally have a good means of relating what they do on a day-by-day basis to these global measurements. Second, there's frequently no incentive for thinking globally. Third, they're not used to thinking in these global terms. Cause-and-effect connections between what they do and the entire organization's well-being are seldom clearly drawn. Furthermore, such measurements take time to get comfortable with, in the same way that kilometers are still uncomfortable for many British people. Management may, through misunderstanding of local measurements or their effects, promote behavior that is inimical to the global goal. Therefore it's important to look at measurements, in particular local performance measurements, in more detail.

Local Measurements

Local measurements serve many purposes. They can

- Allow individuals to evaluate how they are doing, relative (one hopes) to the global objectives. It's always useful to know which tasks are more important, just how important they are, and how each one fits into the overall picture.
- Allow monitoring of project performance and how that performance relates to the global objectives.
- Provide performance feedback for job evaluation. Often some component of bonuses, salary, and/or awards is based on level of performance.
- Provide performance feedback for analysis of whom to promote. Of course, this is very different from routine job performance analysis.

Good job performance doesn't necessarily have anything to do with future performance in another position. The end result of using the same measurement for current performance and anticipated future performance will be the Peter Principle: "In a hierarchy every employee tends to rise to his level of incompetence."[6]

Notice that always the local measurements must be tied back to global performance. Using TOC-based scheduling and the five focusing steps, we can derive some ideas for local performance measurements:

- The strategic resource should be measured on how well it is subordinated to the market. For example, it should not be producing things that aren't needed just because it is strategically considered a leverage point. One useful measurement is throughput (for actual, booked work) across the strategic resource.
- Nonstrategic resources should be measured on how well they subordinate to the strategic resource, if there is one, or to the market, if there isn't. Are they producing what is needed, in the proper sequence?
- Resources should be measured on how quickly they perform their tasks, after the tasks are available to work on (the roadrunner mentality).
- Resources should be measured on the extent to which they use up buffers. If a particular resource is consistently causing problems with buffers, the task estimates might be low, or there might be a large amount of variability in actual task durations. There might also be a bad policy in place, such as multitasking.

Notice that employee suggestions, if useful, will probably fall under an exploit or subordinate rubric. Even cost-cutting suggestions may free up resources that can be used other places to generate more throughput.

Throughput Dollar Days

If a project manager is monitoring how much of the buffer is used up in order to judge the status of the project, shouldn't there be measurements that take that buffer usage into account? The simple answer is "of course." The question is, how should such measurements look?

In a for-profit company, the global measures are financial. In a project, lateness is in terms of time. It makes sense to combine these measurements to estimate the impact of task lateness. Suppose, whenever a task causes consumption of a buffer, we look at throughput times task lateness. For example, if slow completion of a task causes 5 days of additional consumption of the buffer it's feeding, the person responsible for the lateness would be assigned 5 days, times the selling price of the project.[7] Why the entire selling price? Because it is the entire project that is being jeopardized. This measurement is called *"throughput dollar days"* (TDD). It measures what should have been done but wasn't.[8] It quantifies the impact of the lateness. If a late task doesn't cause consumption of the buffer, the lateness has no impact on throughput, and therefore doesn't cause throughput dollar days.

Dollar days should be assigned to the person or people responsible. For example, if you're late because of changed specifications, the reason for the change has to be discovered before the dollar days can be assigned.

This measurement combines two important throughput measures: project throughput and task lateness. The later the project, or the greater the selling price, the greater the impact of lateness. "Negative" dollar days can also be assigned. For example, if you are given work that has already consumed 5 days into the buffer, and when you finish your task the consumption is only 3 days, you have saved 2 days. This is good, and should result in a positive measurement.

Let's look at a few hypothetical examples. Suppose worker X receives some work that has already used 6 days of the buffer being fed. The project selling price or "value" is $10 million. Suppose he then finishes the work 8 days late. He has caused 2 days of lateness, meaning he is assigned 20 million dollar days. On the other hand, if it is determined that he's late because of a mistake of someone upstream, he might be assigned no dollar days. Suppose instead he received the work 1 day early, and after he's done it's 1 day late. He caused 1 day of lateness, which means he is assigned 10 million dollar days.

How should such a measure be applied? It is a very useful way for people or groups to monitor their own performance. People will think hard about ways of avoiding dollar days. If dollar days are assigned appropriately, people are less likely to call their tasks complete before they're really checked out. They know that they'll be assigned the dollar days, probably a lot of them, when the fault is discovered.

Suppose even upper-level managers are assigned dollar days. Such a measurement could reduce the chances of them micromanaging projects with incomplete information, if bad decisions they make would count against them. It would place accountability where it belongs.

Dollar days can readily be used as an informal measure of performance and urgency. Extreme caution should be used if they are to be applied as a formal means of evaluating people's performance. Dollar days are not an objective means of relating local performance to the global goals. The relative importance of projects can't be captured just by the selling price. The mechanism for assigning of dollar days, whether to individuals or to work groups, can be very subjective. And finally, this kind of measurement is very dependent on estimated task durations. On the one hand, this provides additional incentive for people to increase their estimates of task durations. On the other hand, if they don't estimate their own tasks, they are at the mercy of whoever did, and the estimates can be good or bad.

There are various ways that the throughput dollar day measurement can be adjusted. For example, TDD may only mount up after some fraction of of the buffer has been used. This adjustment makes an assumption that some the buffer is expected to be used.

TDD may also be accumulated every day. If we call the total selling price of a project P, the first day someone causes use of the buffer they're assigned P dollar days. The second day they'll have $2 \times P$ dollar days added, the third day $3 \times P$, and so on. This measurement actually increases according to the square of the number of days late, so we could call it TDD-squared. It has the characteristic of putting a great deal of emphasis on lateness. For example, if you have a $10,000 project that you make 10 days late, your total TDD-squared accumulation will be $550,000; a $100,000 project 2 days late will be $300,000. This emphasis can be good or bad, depending on the effects you need to achieve with the measurement.

WIP as a Global Measurement

Work-in-process is easy to measure if it involves investment. That is, if you must purchase something before you can do work, that purchase becomes an investment with financial implications. You can say "I have $20 million in WIP." In many situations the immediate financial picture isn't directly affected by WIP. For example, if an engineer starts designing a new product, the financial ramifications aren't immediately obvious because no money was invested.

1. Throughput

2. Work-in-Process

3. Operating Expense

4. Investment

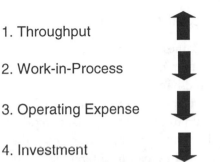

Figure 22-1   Basic Measurements (revised)

Because of its indirect impact on throughput, WIP is strategically important whether or not there's investment money involved. It's so important that traditional TOC, which includes the value of WIP, investment, and inventory together, puts the importance of inventory second only to throughput, and calls it significantly more important than operating expense.[9] This number two status is due to the effect of inventory on throughput, as discussed in Chapter 3. Leaving WIP separate from investment, we might make a more complete list of priorities, with the directions we'd like them to move (Figure 22-1).

If the impact of work-in-process is mainly in terms of time, we should measure it using time. Since its key impact is on competitive edge — that is, the ability to generate throughput — a WIP measurement should also reflect ability to generate throughput. The throughput-related impact of WIP is mainly on lead times of products. Therefore, the best way of measuring it is to evaluate the impact of current WIP on product lead times.

The specific mechanism for measuring WIP will depend on the environment. For example, an organization that produces one main product type may be able to say "our WIP is 3 weeks" and mean that current WIP levels increase the time to get new products to customers by 3 weeks. If products fall into families with similar characteristics, it might make sense to speak of lead time per family. And if there are strategic resources, it will certainly make sense to speak in terms of the lead time at the strategic resources; that is, how far ahead they are committed. It will also make sense to evaluate the impact of new work and new policies in these terms.

Measuring Schedule Changes

Suppose a project becomes high priority. Perhaps something went wrong to delay it, or the customer moved up the due date. Expediting has become necessary. What should happen?

Management must try and move other resources in to complete the project, but they should minimize the impact on other projects. With Critical Chain schedules, they can tell which tasks on other projects are key and should not be disturbed. By creating different test schedules, they can evaluate whether non-Critical Chain resources can be spared, or what is the likely impact of changing their priorities. The impact of expediting can be measured by looking at which projects will likely become late, and by how much. It can also be measured by throughput dollar days: days of lateness times selling price, summed across the affected projects. The biggest impact, the confusion and loss of focus associated with changing priorities, can't be measured directly. It can be minimized using the Critical Chain approach.

Implementing Measurements

It's easy to say "from now on, we'll start calculating and looking at the following measurements." Even if you know the definition, application, and expected effects of the measurements, you need much more to change an organization. While a full discussion of how to make these kinds of changes is beyond the scope of this book, a few comments are in order.

Avoid measurements that pit people against each other. These measurements sometimes seem beneficial to top management; they seem like win–lose to the workers; they are lose–lose for the company. They can build up a climate of noncooperation that can take huge efforts to overcome. For example, suppose teams are measured on who produces more. Will the teams be as likely to share new ideas? Probably not. What would be the effect of a management statement that the lowest-performing 20% of the company will be laid off?

Changing the informal measurements can be a big task. Informal measurements can be communicated with every interaction between a worker and their manager. The managers' attitudes and actions must be in line with the new formal measurements. Some informal measurements also exist as a result of inertia. Frequently when formal measurements are changed, people will fall back on old familiar ones, which have by then become informal. Two measurements we've seen already that can be formal or informal are task completion relative to commitment, and how busy people are.

People must understand the measurements. If they don't understand, or think the measurements don't make sense, anything can happen. Again, "If you measure me in an illogical way ... do not complain about illogical behavior."[10]

People must have a reason to care about their measurements. There are certainly plenty of ways of creating positive and negative reinforcements for behavior. Best is if they care about the organization, and the measurements guide them to do their best for the organization. In such an environment win–win relationships will happen naturally.

Communication about measurements must be two-way. Anyone should be able to offer suggestions. When changing circumstances make measurements obsolete, the first to know will probably be the people who use the measurements daily. If it's not possible to change measurements to improve global performance, people will become much less likely to care about global performance.

Key Concepts

- Tell me how you measure me, and I will tell you how I will behave.
- If you measure me in an illogical way ... do not complain about illogical behavior.
- Local measurements must relate to global measurements.
- Before implementing a measurement, first understand its derivation, application, and effects.
- Throughput dollar days can be useful as an informal local performance measurement.
- WIP is an important global measurement due to its impact on throughput.

Questions for Further Thought

1. What would be some effects of assigning throughput dollar days for tasks that take longer than expected but don't cause buffers to be used?
2. Suppose you decided to measure people based on amount of free time. The more free time they have, the better they are performing. What would be the likely effects of such a measure? Are there other measurements that would be needed?

3. Think of some additional measurements that indicate how well the organization is subordinating to the market, besides due-date performance and work in process.

23 Leveraging the Critical Chain

There are many ways to get more out of the Critical Chain approach; in this chapter we'll look at some ideas and techniques. Many are already in common use. Their applicability will depend on the situation.

The SWAT Team

Many organizations have competent people. Very few have a surplus of really top-notch troubleshooters. Those few people are typically working on the hottest project of the moment. If another project hits a snag that requires them, there are only a few choices:

- Let the project team do the best it can, which may not be good enough;
- Wait until a qualified troubleshooter is available; or
- Pull someone from another project.

An alternative approach is the formation of a SWAT team — one or more people organized to do troubleshooting. With our pit crew mentality we will want to give highest priority to Critical Chain tasks. While SWAT team members may have other duties, their number one priority is to help make sure Critical Chain tasks stay on track. If at any time throughput is jeopardized by a difficult obstacle, they need to be available to help. Their responsibilities may span several projects at once.

This is certainly not a new concept. Frequently technical support people will wear a beeper so that they can be available to solve urgent problems. In some environments that could be all that's needed — personnel with key knowledge being accessible to solve urgent problems or answer critical questions. Sometimes teams will be required. For example, if hardware/software integration is involved, it may be necessary to have key engineers and programmers working together.

Throughput Pricing

What do we need in order to quote a price for a project? Let's consider the case where we have to quote a fixed price for the job. We want to know the minimum we're willing to sell for, in order to decide whether or not to bid. We also want to estimate what is required to win the bid, so that — assuming our minimum price is below that — we can decide how much above the minimum to charge. The traditional approach starts with a rough project plan; resource requirements are estimated; standard costing is applied to get the product cost; and then a margin is added. This cost plus margin is then compared against what is anticipated to win the bid, and possibly adjusted. As we've seen, this cost-based approach doesn't take into account your capacity, and it includes some questionable overhead allocation as well. It is a poor means of pricing projects.

There are two questions that really need to be answered in determining a minimum price: How much is it going to cost to do the project, above current expenses (marginal cost); and how much more could you get by using the resources in another way? The first is usually fairly straightforward to figure out if you have a project plan, because the main components will be material costs, investment costs, and costs of new people. The problem is figuring out the changes to operating expense and throughput that would occur by using your resources in another way. The main avenues would be layoffs and selling other products.

Layoffs will be a big consideration in the Cost World. As we saw in Chapter 15, they will tend to be poorly leveraged relative to increasing throughput, and furthermore you will be getting rid of people, usually your biggest assets. Most likely ex-employees will then try to work for your competition, so the net effect is of giving away your most important assets to your competitors. There are long-term morale and productivity costs to layoffs as well, which should — but too frequently don't — make them a last resort for improving profitability. Yet many companies live and die by this cost-cutting sword. It seems like a curious way to do business.

Using your resources by selling other products can be difficult to evaluate in a project environment, even if there are concrete choices to decide among. Here are some considerations:

- If you have selected strategic resources, you can look at throughput per unit of constraint time, as discussed in Chapter 18. The time required on the strategic resource will give an idea of what the minimum level of throughput should be.
- If you have plenty of capacity to carry out the project, anything you can sell for a price above the truly variable additional costs and should be considered. You have lots of flexibility in winning bids, and should use it to undercut the competition.
- If capacity is short, even if you have selected a control point, consider the effects of adding the project to your existing workload. Will new people have to be hired? Will new equipment or space need to be purchased? Throughput, investment, and operating expense must be taken into account.
- Use standard costing procedures to see what the competition is likely to bid — most companies will look at standard cost plus margin.
- When calculating prices, take into account the competitive advantages you have gained through the use of TOC, including shorter lead times and greater reliability.

Small Batches of Work

As we've seen, work-in-process and lead times have a strong impact on throughput generation. Traditional attitudes commonly create an increase in work-in-process through use of large transfer batches and the use of multi-tasking, as discussed in Chapter 3.

Often tasks are grouped together for some conceptual reason and could in practice be split apart and overlapped. Tasks might typically be grouped according to stage in the development cycle or according to organizational function. If the objective of decreasing lead time is not kept in mind, splitting tasks into smaller pieces may not be considered. Serious attention should be given to whether and how tasks can be overlapped. For example, when creating a new production facility, it might make sense to order equipment before the building is started, or hire and train people before the building is finished. The traditional approach of hiring people and ordering equipment only when there is a place for them to go can lengthen the time until the facility is productive (the project lead time), and thereby lengthen the payback period.

Sometimes several distinct requirements are combined into a single task out of habit. We already discussed splitting formal and informal tests, so that the problems are discovered earlier during the informal tests. This not only has the advantage of moving risk earlier, but also can shorten the formal acceptance process, which is typically at the end of the project and therefore part of the Critical Chain.

Focus on Strategic Resources and Strategic Tasks

A "strategic resource" is typically a leverage point; more production means more throughput. Very often it makes sense to pay extra attention both to such resources and to the tasks they perform. An extra minute of production for such a resource, or a minute saved performing a task, is an extra minute of throughput for the entire organization.

Improvements in projects can be achieved by speeding up Critical Chain tasks. Therefore, the Critical Chain tasks should also be considered of "strategic" importance, and improvements should be sought. Learning how to perform key tasks more quickly will produce benefits on each new project of the same type. Therefore, improvements to Critical Chain tasks are especially important in environments where the same or similar projects are done frequently, such as construction or aircraft overhaul.

Win–Win

Long-term relationships must be win–win relationships or they will not continue to exist. They must benefit all the parties in the relationship. There are many relationships in a business: person-to-person, manager and employee, customer and vendor, and so on. Conflicts often arise when trying to maximize short-term gain at the expense of long-term relationships: emphasis on short-term win–lose vs. long-term win–win. A good test for a performance measurement is whether it fosters win–win behavior among the parties affected by it. Win–win is an important principle to keep in mind.[11]

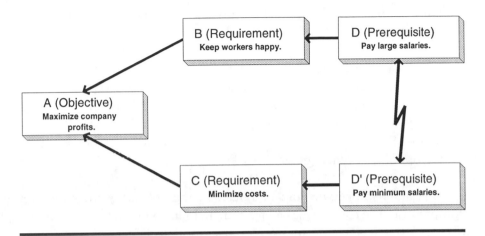

Figure 23-1   Salary Conflict

Table 23-1   Sample Assumptions in Figure 23-1

Arrow	Assumption
A→B	Happy workers are more (globally) productive workers.
B→D	Worker happiness is primarily dependent on salary.
A→C	Cost minimization is a key way of maximizing company profits.
C→D'	Salaries are a large part of controllable costs.
D→D'	The minimum salaries are not considered large.

Let's look at a concrete example: salary conflicts. Workers wish to make as much money as possible; companies wish to make as much money as possible. Furthermore, salaries are measurements. A higher salary implies a "better" evaluation. This means they are causes with possible effects, which means there can be more than just money at stake. Should an employer pay large salaries, minimum salaries, or some compromise?

Figure 23-1 shows the associated conflict or "evaporating cloud" diagram, as in Figure 10-1. The assumptions on all but the D to D' arrow should be read "In order to ⟨x⟩ we must ⟨y⟩ because ⟨assumption⟩." For example, "In order to keep workers happy, we must pay large salaries because worker happiness is primarily dependent on salary." For D to D', read "We can't both pay large salaries and pay minimum salaries because...." Work out some assumptions on your own. Some samples are shown in Table 23-1.

Once we diagram the conflict in this fashion, there is a whole raft of avenues of attack that can be mixed and matched, such as the following:

1. Compromise: a salary somewhere in the middle is picked.
2. A→C: Make monetary rewards dependent on company performance, perhaps through bonuses, profit sharing, or stock options.
3. D→D': Give higher salaries to "better" performers.
4. B→D: Give non-salary perks, such as awards, titles, cars, stock options, or office space.

Item 1 looks like win–lose or lose–lose; this is usually the case with compromise. The company is paying more than it would prefer, the employees are getting less. Item 2, making compensation dependent on global considerations, looks like a real win–win. Item 3 has positive and negative aspects, depending on how "better" is defined. It might pit workers against each other. Item 4 defines the difference between compensation and salary.

Let's consider win–win applied to a single project example, with a very large project. People have been working hard for months. It's really crunch time now, because the project is late. Everyone knows this, everyone is working long hours. And yet....

In a Cost World company, people know that when the project is over they may not have anything to do, which means they may not have a job. They'll put in the hours for many reasons, such as keeping the job for now, looking good, and so on. But probably they won't put in the effort, which means things will still take longer than they need to. The months of hard work have sapped people's enthusiasm, and the finish line doesn't look so attractive.

In a Throughput World company, people aren't afraid they'll lose their jobs when the project is over. They know that management is already looking for opportunities to sell their capabilities. Furthermore, there is likely some kind of reward for completion of the project, in order to add more incentive to complete quickly. This is a win–win environment.

Key Concepts

- Consider a SWAT team if there are key resources needed in many projects.
- Use Throughput Pricing to get more business.

- Use small batches, in order to take advantage of overlapping to speed up projects.
- Focus improvements on strategic resources and strategic tasks.
- Think win–win.

Questions for Further Thought

1. Does a SWAT team imply multitasking? Why, and why not?
2. In what situations might it be bad to split tasks apart so that they can be overlapped in time?
3. In what ways have companies attacked the arrow A→B in Figure 23-1?

24 Weak Links

Very often we get so caught up in the process of implementing or promoting an idea that we lose sight of the fact that how we think about the solution isn't what will cause it to become reality. In fact, how we think about it may be completely irrelevant. We have to keep aware of potential obstacles so that they can be addressed before they block progress.

The NIH Law

Resistance to change is a frequent cause of problems. This is especially true when implementing TOC, because the changes to people's mindsets are often substantial. On the other hand, we have to believe the changes can be made. If we don't believe it, there's no point in bothering to try.

First we need to make a distinction between those who need ownership of a solution and those who need understanding. Understanding of "to what to change" is required for everyone involved with the solution. Ownership is required of those who might have any motivation to block the solution. If you tell the person who mows your lawn to do the back yard first rather than the front, he probably won't care. If you tell him to use a manual mower in the back, you will need to give him a serious incentive.

A solution without ownership is the "Someone Else's Solution" or the "Not Invented Here" (NIH) syndrome. It's virtually impossible to present your solution to someone else without them seeing it as your solution. If it's yours, it's not theirs. They resist it. It was not invented here. There is a formula that describes this resistance:

$$\text{The NIH Law:} \qquad \Omega \propto T/(EC)$$

The variable Ω is a person's resistance to your solution, T is the time they have spent suffering from the problem, E is the esteem in which they hold the presenter of the solution, and C is the complexity of the proposed solution. The resistance you will encounter to a solution is proportional to the time they have spent suffering from the problem, divided by the esteem in which the person holds the presenter of the solution, and also divided by the complexity of the solution. The easier the solution, the longer they've spent suffering from the problem, and the less esteem in which they hold the presenter, the harder it will be for them to accept.[12]

We can validate the formula by looking at what happens in real life. In cases where T is small, we usually have a new person on the job who has not learned to suffer with the problem. Such an individual is willing to consider anything, even a simple solution. He or she comes in with enthusiasm and new ideas and is not loved because of it. The ideas are not accepted; frequently these people will end up losing their steam or leaving. When E is small, there are obvious difficulties that we don't need to belabor. If people really hate you they may even do the opposite of what you suggest, and "reverse psychology" becomes necessary.

What happens when C increases? This is the most unexpected part of the formula. Our instinctive reaction when we meet resistance is to make the solution look simpler. We assume they are too stupid to understand. This is wrong. In fact, with a more complex solution we see resistance go down. Many times we see implementation of solutions that are much more complicated and less effective than they needed to be. We see this in fields as diverse as information systems development and acquisition, health care reimbursement, military contracting — in short, virtually everywhere. We shake our heads and say "why didn't they just...."

Marketing experts will claim that the simpler a solution, the more likely someone is to adopt it. It's true that the NIH law becomes nonlinear and actually reverses with a large C. For a solution to be accepted, someone needs to believe they understand it to some degree. We also agree that the simpler the solution, the more likely it is to work. But if you want to evaluate the NIH formula with an open mind, first pick an example. Ask yourself what solution is being purveyed for what problem, estimate whether E, T, and C are high or low, and see whether the formula says the resistance, Ω, is high or low.

The NIH law is presented as a natural law based on empirical observation, and we avoid claiming it to be good or bad. Indeed, there are likely to be perfectly good reasons for the behavior it implies, such as

- The presenter doesn't fully understand the problem.
- Changes to the system, rather than within the system, are hard to accept because the consequences are difficult to predict.
- People feel secure with established problems (the "better the devil you know" syndrome).
- The person with the problem is unwilling to look stupid by admitting there's a simple solution to a chronic problem.

There are also bad reasons that we don't need to go into.

What can you do if you have a simple solution for someone else's problem? There are obvious answers, such as making the solution complicated (C goes up), having someone they respect more present it (E goes up), or presenting it as a solution for an entirely different problem (T goes down). It's easy to find examples of each. These are tried and true approaches, not to be shunned, but there is another approach that we also recommend considering.

Your solution must become their solution. They must embrace it, which requires that they have a part in developing it. If there are many people affected by a problem, most of those people, and all those who can block the solution, must have a say in developing it. Practical managers can be happy because the ultimate solution may be improved by this process. Theoreticians don't worry about that, and merely observe that both E and C go up.

An interesting paradox accompanies this process. We usually look for an open mind in those we wish to convince. Unexpectedly, it is far more important for the presenter of new concepts to have an open mind than those being presented to. The presenter must understand that there won't be a solution to anything until they get consensus. Real consensus almost always implies that the proposed solution is changed. It may not get better with the changes; almost certainly it will get more complex. The presenter must therefore be careful of two things: that people take ownership over the solution, and that the quality of the solution suffers as little as possible during the process. Sometimes speed and efficiency must be sacrificed.

If the people responsible for solutions do not have responsibility for the problems, they will face another obstacle in promoting their ideas, which is sometimes termed "Someone Else's Problem" (SEP). They may not have a really strong incentive to get things done, and the people who own the problem don't expect them to. A simple example is the Information Technology department whose measurements are divorced from a global viewpoint. If this department were to announce to a company "we have a Critical Chain scheduling program, please come use it," there's a good chance that they won't — and possibly can't — put forward the resources actually to educate people and implement the software. They can suggest the solution, they can even promote it, but they can't be the ones to make it happen.

SEP is similar to NIH, but worse. The major differences are twofold. First, solutions are even less likely to be accepted (the value of E, esteem, is low), which means that the people with actual responsibility must have even more to do with developing them. Second, the people presenting the solutions are likely to think something like "it's not our problem if you're too stupid to understand this" and abandon further efforts.

Boundless Optimism

Some people are so excited by a solution or approach that they immediately plunge into it full steam. This kind of passion can be very powerful and very productive. It can also mean you're going 90 miles per hour when you hit a brick wall.

If you're a very enthusiastic type of person, consider some suggestions:

- Take time to plan. Who needs to be involved? What information are you missing?
- Consider the negatives. In fact, solicit negatives. If you don't, they may bite you. Even if you know they wouldn't, someone will bite you because you haven't considered them. (The opposite syndrome, "unguarded pessimism," makes everything into a negative. If you have this problem, you'll probably be regarded as a walking pothole, and people won't accept your ideas anyway.)
- Don't ignore signs of trouble unless you expected them and planned for them.
- Team up with someone who has a more sober outlook.

In short, make sure the pool is full before you dive in.

Milestones

As we saw in Part II, milestones (also called intermediate due dates) that must be met can have the effect of increasing the required time for a project, because each must be protected with extra time in the schedule. What can you do instead? A given task's completion time is really only important relative to its impact on the buffers it feeds. That implies that milestones, if necessary for tracking projects, should be expressed in terms of the buffer, as with the visit to Emily in Chapter 8. We expect tasks to be completed within some window of time. The early time represents no buffer penetration; the later time represents using up the entire buffer. Actions taken will depend on the amount of buffer remaining, the relative duration of tasks remaining, and the relative risk remaining. If there's lots of risk and not much buffer, drastic actions may be required. It's worth experimenting with ways buffer reporting can be incorporated into projects for both internal and external control and reporting.

Good Enough

How good is good? There's a common tendency, especially when people can't see the global picture, to want things to be "perfect." Perfect means as good as possible, according to some ideal view of the world. Programmers want the code to be just right, engineers want the product to be elegant. People need to understand "good enough" in the context of any project. They need to understand the project requirements well enough to know what is and isn't required. Overdesign, building to unnecessary tolerances, and focus on unnecessary capabilities have all killed projects. Management needs to keep everyone's eyes on the ball.

Figure 24-1 shows what usually determines "good enough." The customer might be an external client or a group or individual internal to the company. The closer these viewpoints can mesh, the happier the customer, the happier the developer, and ultimately the better both will achieve their goals. Any work that doesn't contribute to this focusing, now or in the future, is beyond "good enough." The arrow between customer and developer represents a communication channel that must exist for the viewpoints to be synchro-nized. If there's a debate over what's good enough, ask the customer. If you can't ask the customer, the structure of the project may be flawed.

When implementing Critical Chain concepts, you need to take this a step further; the motto becomes "good enough for now." Some improvement is better than none. Far better to make a little progress each day than to try to make it all and fail miserably. It's tiresome but useful to recall the old joke, "How do you eat an elephant? One bite at a time." How do you get other people to eat an elephant? You'd better make sure both the people and the elephant are properly prepared, otherwise save a lot of space for leftovers.

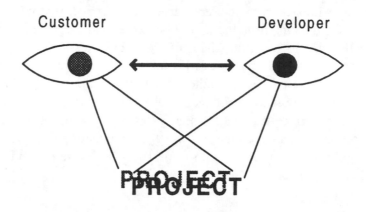

Figure 24-1   Project Focus

The Cost World

Consider a scenario: Your project is going well. You're ahead of schedule, and hope to finish early and under budget. Risk seems to be under control. Then, over the course of a few months, resources are taken away. Schedules slip, your slack is used up, soon on-time completion of the project itself is in jeopardy. What happened?

It could be that by managing the project well, you have exposed significant extra capacity. You have resources that don't have to be kept busy on your project, which means they can be used "more efficiently" elsewhere. It could also be that your contract has incentives to minimize costs in any given time period, which make it seem even more desirable to cut costs by taking away resources.

Both these reasons are the result of Cost World thinking. Both result in local optima that almost certainly did not contribute to the global optimum. Both can be mandated from above without your input. They can easily kill the pit crew mentality; it's hard to give a project your best if you know that you or the project will suffer for it.

"Management" commonly means dealing as effectively as possible with the things you have. It tends to mean dealing with links rather than chains. It often implies an in-the-box focus. A common philosophy among managers is that of Lord Chesterfield: "Take care of the pence, and the pounds will take care of themselves." This is the Cost World point of view. Consider this quote:

> If The Entrepreneur lives in the future, The Manager lives in the past.
> Where The Entrepreneur craves control, The Manager craves order.
> Where The Entrepreneur thrives on change, The Manager compulsively clings to the status quo.
> Where The Entrepreneur invariably sees the opportunity in events, The Manager invariably sees the problems.[13]

The Throughput World implies looking out of the box. "Leadership" and "entrepreneurship" are related terms that imply looking out of the box. This quote expresses a common viewpoint. It states a conflict between being a manager and being an entrepreneur; we can derive from it an implied conflict between being a manager and working in the Throughput World.

There need be no conflict. No doubt managers must manage. In order to manage successfully in the Throughput World, they must also be leaders; they must be entrepreneurs. They must plan for the future while taking care of the present. They must have the organizational support to make those attitudes work.

Leverage Outside the Box

We have been advocating an "outside-the-box" focus; the Throughput World seems to demand that. On the other hand, leverage points are often inside the box, and require that significant attention be paid there. On the surface this seems to be just an irrelevant philosophical point. In fact it can be a serious contradiction that must be understood before implementing any improvement process.

It's not uncommon for companies to institute successful improvement projects, especially if they believe in the existence of leverage points. Suppose, for example, that a printing company improves their graphics design department — historically, the constraining resource — to the extent that there is clearly extra capacity. Now the company is looking to translate this improvement into bottom-line results. What are the options? One option, often the simplest and quickest, is to cut costs; that is, lay people off. Where will they be laid off? The logical places would be where there is extra capacity; in this case, definitely at least in the graphics design department. So the "reward" for improvement is to be laid off. The bigger the improvement, the bigger the reward. The cause and effect are clear to everyone in the company. What is the impact on morale? What is the chance of success of future improvement projects in graphics design, or for that matter elsewhere in the company?

Cost cutting is frequently the path chosen. It is the "in-the-box" approach. There is usually a much better way, but that way requires an out-of-the-box mindset; it requires looking at the organization as a chain whose links must work together. Increase throughput. Get more sales, produce more. Such an approach requires a more integrated approach to managing the company, both internally, in terms of coordinating functions, and externally, considering its relationship with the rest of the world. It requires a shift from the Cost World to the Throughput World.

Education

Several times we've mentioned the need for education, and we'll mention it once again. Education, properly done, is a means for people to take ownership over a solution. Using the NIH Law terminology, E (esteem) goes up. The actual complexity of the solution C doesn't really go down, so the net effect is positive. People must understand specifically what you're trying to do, how it affects them, and logically how the solution relates to the goal of the organization. If they don't understand what you're trying to do or how it affects them, they'll have to like you a lot before they'll help you to do it. Even then they won't get it right. If they don't understand how the solution relates to the goal of the organization, they will be less likely to contribute effectively to it. With proper education, people are more likely to agree that a concept really can become a solution to their problems.

What is "proper" education? Again, we refer back to the NIH Law. "It is far more important for the presenter of new concepts to have an open mind than those being presented to." Proper education requires the educator to work with the students in fitting the concepts into the students' reality.

Key Concepts

- The NIH Law, $\Omega \propto T/(EC)$, expresses resistance to change.
- It is even more important to have an open mind when presenting a solution than when having one presented.
- Make sure the pool is full before you dive in.
- Avoid intermediate due dates that aren't expressed in terms of buffers.

- Understand the meaning of "good enough."
- Managing in the Throughput World implies both management and leadership.
- Reward workers with layoffs and they will reward you with lack of enthusiasm.
- Promote understanding of the changes you're trying to make.

Questions for Further Thought

1. In military organizations, officers are restationed every few years, while civilian managers might hold a position in the same group for many years. Based on the NIH Law, which type of person would on average be likely to have the least resistance to new solutions to work-related problems?
2. If you want someone to eat a whole elephant, at what point should you tell them what they're eating?
3. What is the relationship between "good enough" and Parkinson's Law?

25 Implementation Checklist

Now that we've gone through many TOC and Critical Chain principles and techniques, let's summarize them and how they apply to you. Rather than repeat what we've already covered, let's instead think in terms of a set of questions. These are the kinds of questions you will need to ask in order to apply TOC in an environment you've chosen. Some questions will be simple, but not all questions can be answered right away. All of them should be considered; not all will be relevant.

Take Stock

Do you believe you've already implemented the concepts in this book? If so, test your belief. Pick five people in your organization at random. Ask each to state the goal of the organization, and where the key leverage points are. Were they able to answer? Did the answers agree? If the answer to either question is "no," then Theory of Constraints is not fully implemented in your organization.

What Is Your System?

You must describe the system you're trying to improve. What is it? What does its box look like? You can look at the system with an external or an internal viewpoint; that is, from the outside or the inside.

Externally, some aspects of the box are the products you make and the markets you go after. How does your system relate to other systems? Are you in the middle of a chain of vendors and customers? Are your markets highly competitive, and is the technology constantly changing? It's important to look often at the external picture, climbing out of the box to see where the walls are and what they're made of.

You can also analyze your system internally, looking at how its components fit together to make the whole. Sometimes it's useful to diagram how products flow through the system. It can also be interesting to see how information flows through the system, both through measurements and through direct communication. Does the marketing department base any of its decisions on available resources? How do the salespeople tell if projects are going to be late? How do people planning a project find out that it's all right for the project to be late?

What Are Your System's Goal and Necessary Conditions?

The importance of understanding the goal can't be stressed enough. The goal will lead to a measure of throughput, and therefore a sense of where everyone's actions are pointed. The goal must come before a vision, because a vision statement usually describes not only the goal but ways in which it will be achieved. Without a goal, without a definition of what the owners of a system want from it, a vision is meaningless. The system-level viewpoint always comes back to the goal.

Necessary conditions are important to identify, especially in not-for-profit organizations. Sufficient cash is usually a necessary condition. Expenses might constitute a necessary condition if, for example, funding levels are fixed. There might be necessary conditions dictated by the company's charter, or by environmental or moral concerns. Some necessary conditions will depend on what the owners of the system want.

What Are the Implications of the Throughput World?

Let's assume you have defined throughput in your environment. Take a few moments to think about the implications of a Throughput World mentality on topics such as the following:

- Your mission statement: does it remain the same?
- Your vision statement: is it in line with your goal?
- Your current improvement projects: are they geared toward cutting costs, or generating more throughput?
- Your customers: are you promoting the right products? Are you pricing products based on your costs plus a margin, or are you using throughput pricing?
- Your vendors: are your purchasing policies in line with saving money, or supporting generation of throughput?
- Budgeting: are people penalized for not spending enough money, even if the extra money spent would generate no additional throughput? Are they penalized for spending too much, even if the additional throughput is significant?

The Throughput World is the out-of-the box, unlimited horizons viewpoint. What is the potential for your organization?

Where Are the Leverage Points?

A belief in and understanding of leverage points is crucial to major improvements. There are several types of leverage points.

Physical Leverage Points

This category includes both the Critical Chain itself and specific strategic resources. The Critical Chain can be identified and buffered using the six-step approach given in Chapters 11 and 12. Be careful — there may be resource contention between projects. If projects share tasks or resources, they can have an impact on one another's schedules. Scheduling them in isolation may produce only a local optimum. This was discussed in Chapters 19 and 20.

If a strategic resource has been selected, the company has decided to regard that resource as important. One job of management is to make sure that other resources don't get in its way. In other words, management must make sure that the selected resource remains the leverage point.

If on the other hand a constraining resource was not selected but has been identified, it should be eliminated as a constraint as soon as possible.

This might be done by hiring or buying more of the resource, by removing the need for the resource, or by finding better ways of utilizing what is available.

Market Leverage Points

Commitments made to customers are extremely important for most companies, both for current and future throughput. If customers receives a poor quality product, if they receive the wrong product, if they receive it late, there's a good chance they'll look to a competitor for future business. Satisfied customers are a necessary way of protecting future throughput.

This is why the market is always considered to be a leverage point. The Critical Chain is scheduled from future to past in order to avoid jeopardizing customer commitments as far as possible. "What-if" analysis and Critical Chain scheduling should be used to determine realistic commitment dates, in order to decrease the chance that customers will be disappointed. Critical Chain scheduling and buffer management should be used to improve reliability, and if necessary inform customers well in advance of problems with their delivery.

Which Policies Are Constraints?

Most environments have policy constraints, which are existing policies, formal and informal, that inhibit the generation of better bottom-line results. It's not enough to say that a policy is bad without being sure a better one exists — a revised policy that will improve the results of the organization. Therefore, in order to qualify as constraints, policies must be associated with one or more injections that will improve performance.

Sometimes policies can just be reversed. Usually much more is required. For example, efficiency measurements are a policy that can constrain an organization, because they cause people to want to look busy. Just taking away formal efficiency measurements is usually not sufficient to improve. In fact, a lack of formal measurements could have the effect of making things worse if people don't know what to do. Better measurements must be provided; people must be educated regarding their meaning and applicability; and the entire organization must act in accord with these "better" measurements. Often many injections are required to fix policy constraints.

Many questionable policies have been mentioned in this book. Policy constraints can sometimes be identified by prior experience. However, people and circumstance are infinitely inventive, and there exist far more constraining policies than could be discussed here. The Current Reality Tree process is designed to help identify problematic policies by showing the causal links between core problems and their associated undesirable effects. Here are some statements that may hint at policy constraints:

- We can't sell below cost.
- It makes no sense to hire someone if they won't even be half utilized.
- Everyone should have enough work on their desks to keep busy.
- People who are overloaded with work are more valuable.
- We won't start anything we don't intend to finish.
- Get rid of the buffers; we don't plan to fail.
- We just want employees to do what they're told.
- Follow the schedules exactly.

How Can the Five-Step Improvement Process Be Applied?

A thorough understanding of the five steps is important in establishing ongoing improvement in a company. To recap, these steps are as follows:

- **Select/Identify** the leverage point(s). The leverage points might be resources, policies, or the Critical Chain. Strategic leverage points (including the market and/or resources) should be selected; others should be identified. Identification can be done using the Current Reality Tree, Critical Chain scheduling, or intuition; selection should be analyzed from a business-wide perspective.
- **Exploit** the leverage point(s). Do your best to squeeze more from them. If the leverage point is a policy you can't change, you may have to do your best given the current policies. Scheduling is a good starting point for exploiting physical constraints.
- **Subordinate** everything else to the above decisions. Subordination will be based on the results of the previous two steps. This step will help subordinate the entire organization to its goal. Placement of buffers and implementation of buffer management are important aids for subordination. Communication of how people relate to the leverage

point(s) and the goal, including various types of measurements, is essential. Sometimes an inability to exploit or subordinate to leverage points can help point out erroneous policies.

- **Elevate** the leverage point(s). If it makes sense (see Step 5), take a significant action to get more of the leverage point. This action will require some operating expense and/or investment. If the leverage point is a policy, political capital may need to be invested.

- Before making any significant changes, **evaluate** whether the leverage point(s) will and should stay the same. Any policies you plan on changing, erroneous or not, should be evaluated to check the likely effects of removing or replacing them. Some evaluation decisions will involve tradeoffs between measurements, for example more throughput for more operating expense. Evaluation can be done using Critical Chain scheduling and/or the Future Reality Tree.

Where Does It Make Sense to Start?

Dettmer[14] presents the concept of "sphere of influence," those things over which you have some influence, vs. "span of control," which includes the things over which you have control. Obviously, the distinction won't always be clear-cut, but the less influence you have over an area, the less directly you'll be able to cause changes.

Explicitly select areas in which to start implementing the TOC concepts; don't expect to do everything at once. Plan it out. What positive effects do you hope to achieve? Very often a successful first attempt, even if not highly leveraged in global terms, will give you the influence to expand into more useful areas. Starting to implement TOC might involve giving someone a book, expediting a late project, or scheduling a new project. But take care that the implementors aren't viewed as the "owners" of the concepts, or the NIH law may come into play.

Which People Will Block Implementation?

This implies a number of questions:

1. *Who are the people who will be involved, and how will they be involved?* Depending on where you start there will be workers, higher-level management, resource managers and project managers.

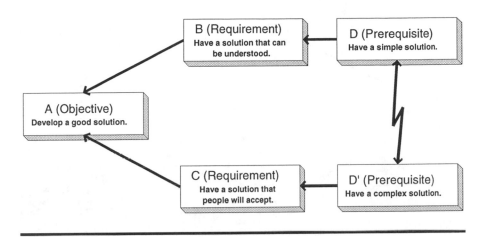

Figure 25-1   Simple Solutions (Conflict)

2. *Do those involved just need understanding, or ownership?* This will help determine who needs to be involved in developing the solution.

3. *Who will be threatened by the solution? Do you care?* Very often people find change threatening. Sometimes it doesn't matter; sometimes you will need to take that into account, and make sure the win–win is clear.

4. *Could the changes continue after you're gone?* Very often TOC implementations are the results of one or two "champions" who are responsible for applying the concepts in a company. Sometimes this is inevitable at the beginning, but for the long term the process of change becomes very vulnerable as people leave.

5. *Are you blocking the implementation?* Enthusiasm without others taking ownership will make change difficult. If everyone agrees that changes are logical and needed, yet they don't happen, it means that they don't have ownership (which may also mean they don't understand). Keep an open mind when presenting your ideas.

6. *Is the solution simple?* Here we have an interesting conflict. In general, a solution should be simple, or else it can't be communicated and implemented. On the other hand, based on the NIH Law a certain amount of complexity will arise from the process of allowing people to "improve" the solution. Figure 25-1 shows the conflict, expressed in a diagram. Before reading on, study Figure 25-1 and think about how the conflict can be resolved. Surface some assumptions behind the arrows. If you are tempted to compromise, think again.

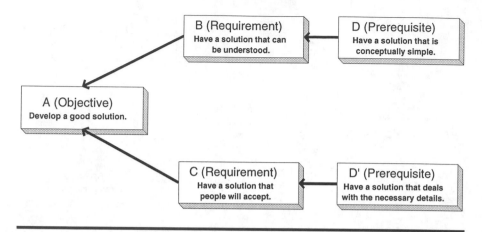

Figure 25-2 Simple Solutions (Resolution)

An assumption between D and D′ is that a solution can't be both simple and complex. In general that's not true. There can be solutions that are conceptually simple, but complex at a detailed implementation level. Consider, for example, a nuclear disarmament treaty, where absolute buy-in is required from all parties before they'll sign. The solution can be communicated and accepted, even though the details may be extremely complex. Therefore, one resolution of the conflict might be Figure 25-2.

What Is the Status of the Leverage Points and Their Buffers?

If we can assume that the leverage points are key for gaining more throughput, and that buffers are key in subordinating to leverage points, then it makes sense that the status of leverage points and buffers will be well known in the organization. Very often this is not the case, even in organizations where there is a great deal of TOC understanding. Monitoring the leverage points and their buffers is exactly like keeping your eye on the ball in baseball.

How Do Causes Relate to Effects in Your Environment?

Try to understand your environment in terms of cause and effect. Working through the existing causalities and the expected results of actions can pay big dividends. The Current and Future Reality Trees are excellent techniques for doing this. Some references have already been given. Practice explicit effect–cause–effect thinking. Study other approaches, such as fishbone (Ishikawa) diagrams or "systems dynamics" as pioneered by Dr. Jay Forrester,[15] and identify their strengths and weaknesses. Logical thinking can be learned, and it is much more effective in achieving goals than illogical thinking.

Key Concepts

- Describe your system.
- Verbalize your system's goal and necessary conditions.
- Imagine in what ways the organization would look different if it were in the Throughput World.
- Examine the organization with the point of view that there are leverage points.
- Look for people and policies that can block you.
- Keep the solution as simple as possible, especially at a conceptual level.

Questions for Further Thought

1. What do you still need in order to believe that you can cause dramatic improvements (Miracle 4)?

Endnotes

1. For a discussion relating the five-step process and the three questions, see Goldratt, E. M., *What is This Thing Called Theory of Constraints and How Should It Be Implemented*, North River Press, Croton-on-Hudson, NY, 1990, 7–8.
2. William Shakespeare, *The Rape of Lucrece*, line 353.
3. In addition to other books already mentioned, see for example James Cox and Archie Lockamy, *Reengineering Performance Measurement*, Richard D. Irwin, New York, 1994; Mokshagundam Srikanth and Scott Robertson, *Measurements for Effective Decision Making*, The Spectrum, Wallingford, CT, 1995.
4. Dr. Eliyahu Goldratt, *The Theory of Constraints Journal*, New Haven: The Avraham Y. Goldratt Institute, 1987, Vol. 1, No. 3, 1.
5. Goldratt, E. M., *The Haystack Syndrome*, 28.
6. Dr. Laurence J. Peter and Raymond Hull, *The Peter Principle*, William Morrow, New York, 1969.
7. Projects don't always have a selling price, as for example with new product development. In that case a number must be created to represent the expected throughput, such as expected return within a year.
8. Dollar day measurements are discussed in *The Theory of Constraints Journal*, Vol. 1, No. 3, 17–21; and in *The Haystack Syndrome*, 146–155.
9. For more discussion of this order of importance, see Goldratt, E. M., *The Haystack Syndrome*, 47–51.
10. Goldratt, E. M., *The Haystack Syndrome*, 26.
11. See also Stephen Covey, *The Seven Habits of Highly Effective People*, Fireside, New York, 1990, Habit 4.
12. We can argue whether, or how, the quality of the solution should be incorporated in the formula. Further research is needed. For now we assume that the proposed solutions are good enough.
13. Michael E. Gerber, *The E-Myth Revisited*, Harper Business, New York, 1995, 25.
14. Dettmer, H. W., *Goldratt's Theory of Constraints*, ASQC Quality Press, Milwaukee, WI, 1997, 67–69.
15. For a good introduction, see Nancy Roberts, David Andersen, Ralph Deal, Michael Garet, and William Shaffer, *Introduction to Computer Simulation*, Productivity Press, Portland, OR, 1994. There are many other references.

Afterword

We have reached the end of the book, but we're still at the beginning of the journey. Theory of Constraints itself has come a long way, and still has a long way to go.

In the Introduction we talked about how far TOC has come. Many of its fundamental principles were viewed 10 years ago as revolutionary, even crazy; now they are taught in many major corporations and universities. TOC and TOC-related concepts have become competitive weapons.

How much farther is there to go? Consider the statement made in the Introduction: "It is not uncommon for the application of TOC to reduce factory inventory levels by 75% or more and increase throughput by 40%. These dramatic improvements frequently come after more conventional approaches have failed." Theory of Constraints can frequently help companies increase throughput by 40% or more, without significantly increasing operating expenses. Even if you choose to accept a quarter of this figure, 10% is still significantly more than other improvement initiatives can consistently boast. Furthermore, most conventional techniques depend on cost-cutting, which provides short-term improvements to the bottom line while often demolishing short-term throughput and long-term competitiveness.

Now try to estimate how many companies are actually using the full leverage of Theory of Constraints. Apply the test at the start of Chapter 25 to your company, have friends and colleagues apply the test, and get a sense of how many companies have really implemented Theory of Constraints. The answer most people will arrive at today is "very few." We have a proven, common-sense methodology that demonstrably produces results; it should be part of every organization and business curriculum in the world. It isn't. Why not?

One reason is that Theory of Constraints is perceived, and often pre-sented, as requiring discontinuous changes to an organization.* It may pro-vide a quantum leap in performance; in many cases that leap necessitates significant organizational change. Only truly visionary or truly desperate companies are willing to bet the farm on something that isn't already part of common practice. We can validate this hypothesis by observing that many companies have visionary individuals who wish to apply TOC; but during the implementation process the TOC concepts typically become much more mainstream. It's worth noting that even these diluted implementations still produce significant benefits. Meanwhile downsizing, which is usually far more destabilizing than TOC, seems like a continuous change, because we understand before we start just what we'll end up with. Or so we believe.

Another obstacle to the widespread acceptance of TOC is lack of a good definition of the "whole product." There is no standard, soup-to-nuts imple-mentation, and nothing that says what it should be. You can buy a project scheduling system and understand what you're getting and what you're pay-ing before the system is purchased. You'll know that any problems have already been experienced by many others, and that there are consultants or books that will have the solutions. You can't be sure this will be the case with TOC. We're buying TOC; what's that? Does it blow up when you pull the trigger? A visionary CEO who wants to implement it must have faith in her understanding of the concepts and faith in her organization's ability to assim-ilate them. This is uncommon.

These obstacles will be overcome, if only because market pressures demand it. The improvements are there to be made. How can the implemen-tation process be made as smooth as possible? What is the whole product being implemented? These questions must be answered each time an indi-vidual or company or consultant tries to implement TOC. We have supplied some ideas in this book, but reality is infinitely inventive. There aren't often easy answers, and there will be no answers that fit all circumstances. It's clear that, as the world opens up to TOC, TOC will inevitably open up to the world. Sometimes power and logic will be sacrificed to practicality. Undoubt-edly the path ahead is long and sometimes difficult. It is also immensely rewarding.

---

* For a very useful and readable marketing book that deals with this subject, read Geoffrey A. Moore, *Crossing the Chasm*, Harper Business, New York, 1991.

APPENDICES

Appendix A:
Answers for Further Thought

The "Questions for Further Thought" are designed to provoke critical thinking about issues presented. Most of them do not have one "correct" answer. Our intent in this section is to provide some useful ideas and stimulate further thought, not to provide definitive answers. We recommend that you think carefully about the questions before consulting this section. After reading the comments here, think some more.

Introduction

2. We don't want to be too pompous about this. But whether or not you are "right" has to be evaluated based on achieving real improvements, not on how well any book is followed. Furthermore, what is right today may be wrong tomorrow. Many of the ideas in this book may lead you to the right solutions to some of your problems, but they do not absolve you from thinking those answers through.

Chapter 1

2. Based on the material covered so far, it looks as if a historical database giving durations of various kinds of tasks could be reasonably useful. After all, part of the problem seems to be knowing how long the tasks will take. If we could compare tasks with the historical experience, we could apparently obtain some answers.

3. Uncertainty means we don't know what is going to happen. In general, the preferred methods of dealing with uncertainty are to eliminate it by converting uncertainty to certainty (that is, get more information), to eliminate it by removing the causes, and to protect against it, by adding padding (a "paranoia" factor). Eliminating the uncertainty must usually be done at a per-task level. "Paranoia" is usually considered at a per-task level as well, although (as we'll see) it doesn't have to be.

In order to convert uncertainty to certainty, we typically make more detailed specifications, provide more detailed supervision, or collect more detailed historical data. These can sometimes help convert some uncertainty to certainty. In order to remove the causes of uncertainty, we might try to control the processes better. This can be done by using better quality control, whether we're dealing with computer programming or chemical processes. However, we can never eliminate all the uncertainty, which is why a paranoia factor must be added.

When a schedule becomes invalid due to uncertainty, new schedules are created. As a consequence, in many environments new schedules must be generated daily or weekly, even for projects that last months or years.

If the existing approaches to dealing with uncertainty worked, we wouldn't have to deal with the problem here. The reality is, we can't precisely predict the future, no matter how hard we try. Some sources of uncertainty cannot be eliminated. Therefore, we must acknowledge uncertainty, and we must have means of protecting ourselves from it. The Theory of Constraints concept of the "buffer," discussed in Part II, is a useful means of protecting projects, rather than tasks, from uncertainty.

Chapter 2

1. We should only care if tasks are late when they make projects late, or when late tasks take away resources that would allow us to complete more projects.
2. Task duration predictions are self-fulfilling prophecies whenever they are taken as facts by people doing the work. Most of the time there are negatives associated with finishing a task late, but little benefit to finishing a task early.

3. The question does not indicate "undesirable to whom." Assuming we mean undesirable to Joe or Paul, here are some statements they made that may indicate undesirable things.
   - Joe: I kind of wish I had more work.
   - Paul: We're over budget on most of our projects.
   - Paul: We'll wait until it's too late and then ask for help.
   - Paul: Nobody has the slightest clue how long anything is really going to take.
   - Joe: The powers on high usually cut time from our estimates so they can make competitive bids to get new work.
   - Paul: The design work we get from you guys has — uh — occasional mistakes.
   - Joe: Remember that layoff last year?
4. All of these undesirable things are related, therefore the solutions must be related as well. Chapters 5 and 6 discuss how they relate.
5. No.
7. There are some questionable policies at work in Joe and Paul's environment, policies that cause people to take on multiple tasks, stay busy, and so on. Assuming we're looking for real improvements, these policies will probably have to change. If these policies change, any historical data based on the original policies will be flawed. Historical data will only be of value if Televar does not plan on making significant changes.
8. The resource is not doing anything (it's available). There is work the resource could do that would improve the project's situation. Ordinarily, it could only be good to add the resource. However, there are contracts for which adding the resource would not materially affect the company in a positive way. In some contracts the time spent by existing resources is paid for, the time additional resources would spend is not paid for (or not paid as much), and the late project doesn't result in a significant penalty. Even so, there can be long-term benefits to completing on time and having satisfied customers.

Chapter 3

1. This time may not be a reasonable measurement of your work in process, if you've included tasks you don't intend to complete.

2. Most people experience significant buildups of WIP in areas such as yard work, reading material, and personal correspondence. Often this results in batching (lead times are longer) and multitasking (everything is delayed). Using WIP as a personal measurement has implications both on immediate actions (such as accepting fewer WIP-increasing tasks) and benefits (such as being able to complete important tasks more quickly).

3. Joe gives high estimates for his tasks. This by itself increases work-in-process by increasing the amount of time the work will likely stay in the system. It also encourages multitasking, which increases work in process still further. Lead times in such an environment are likely to be much longer than the actual time to do the work. That is, work tends to sit around for long periods waiting to be completed.

Chapter 4

1. Procrastination can arise because it's undesirable to work on a task, or because it's undesirable to complete the task. Parkinson's Law can manifest itself as procrastination when it's undesirable to complete the task. If you want to postpone finishing a task in order to look busy, you may procrastinate and work on other things.

3. Frequent setting up means frequent changing from job to job. Many errors are possible due to not having the context of the new task firmly in mind.

5. You can't start a task early if the person working on it before you hasn't finished early. According to Parkinson's Law, that person is unlikely to finish early even if he or she could; therefore you are unlikely to be able to start early.

6. Task start times are necessary for tasks with no predecessors. Otherwise start and finish times can be self-fulfilling prophecies that cause delays. This subject is discussed in detail in Chapter 21.

Chapter 5

1. Many times contracts, especially military contracts, are specified as cost-plus. The contractor is paid some amount to cover their "costs,"

plus some amount as additional profit. In such cases it's not uncommon for people to be dedicated to particular tasks or projects. This means multitasking can be minimal. Due to the nature of these contracts, people's time is paid for based on time spent rather than work accomplished. This means there will be a strong tendency for work to expand to fill the time over which the workers will be paid, whether or not they're doing anything productive on the project. In other words, there is a form of Parkinson's Law, reinforced by being paid for the time spent.

2. In this case, the salespeople function like the project managers in Figure 5-2. If you substitute "salespeople" for "project managers" in boxes 9 and 10, you will get one possible scenario. Probably you will have to add the entity "workers prefer to hear fewer screams from salespeople."

4. There are three main loops: 8 to 16 to 18 to 11, 8 to 17 to 19 to 20, and 8 to 16 to 23. These feedback loops ensure that estimation of available resources vs. required resources will be pretty much impossible. They ensure box 8. It's possible to draw other loops, as can be seen in Chapter 6.

5. Figure A-1 gives one possible example.

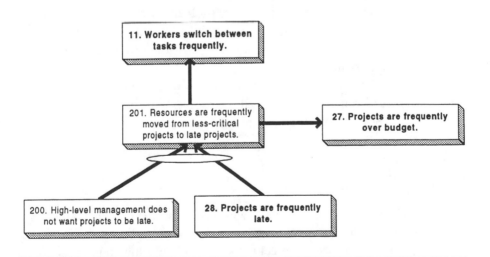

Figure A-1 Moving Resources

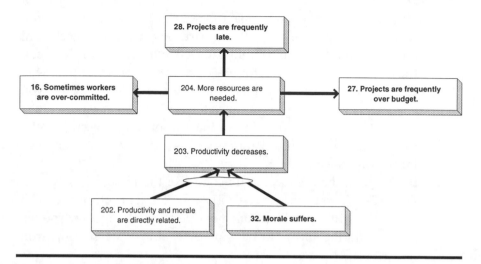

Figure A-2   Productivity Decreases

6. Figure A-2 gives an example of how this can fit in. "Productivity decreases" can result in several important loops that have very negative effects on project performance.

Chapter 6

1. People need to understand that the problems are shared by everyone, and that win–win solutions are possible. In other words, they need to adopt a mentality of "you and I against the problem" rather than "you against me."
2. The entity "Sometimes internal rules prohibit the use of available resources" can serve to make all the problems at the top of Figure 6-1 worse. See, for example, Figure A-3.

Chapter 8

1. It looks as though Jack's tasks will now be on the Critical Chain, and consequently Janet's will need a buffer. Working backward from the time the lesson starts, we get the kind of picture shown in Figure A-4.

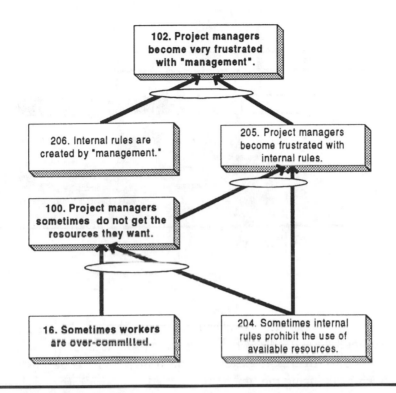

Figure A-3   Internal Rules

The Critical Chain, composed of tasks with bold outlines, has changed. Why do Janet's tasks need a resource rather than a feeding buffer? Janet's errands are not prerequisites for the trip to karate. As far as the trip goes, she could do the errands any time. The car is definitely needed. In order that the car resource can be available on time for the trip, we have put in a resource buffer.

2. No, there would be too much padding in the schedule. She would have both padding to protect the tasks, and buffers to protect the project.

3. Janet cared about the completion time of the trip to karate; that is, the project completion. The schedule was created in order to protect that time. In a global sense, other task completion times were irrelevant, as long as the tasks were completed and didn't make the project late.

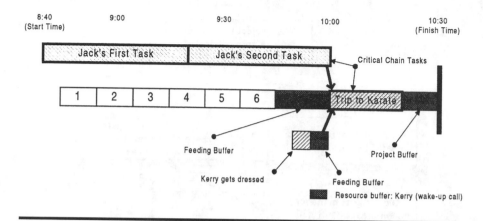

Figure A-4   More Work for Jack

Chapter 9

1. Shorter buffers mean smaller lead times. This can be a significant competitive advantage when bidding for projects. Shorter buffers also allow investments (e.g., for outside contractors) to be postponed, thereby reducing the time money is tied up and improving the cash position. Realistically, in order to shorten buffers one must either decrease the variability in task durations, or accept a reduced probability of on-time completion. Variability in task durations can be reduced in a number of ways, depending on the tasks. See also the answer to Question 3 from Chapter 1.

2. Longer buffers imply more protection; this protection helps ensure that the projects complete on time. Buffers are the "paranoia factor." Everyone who sees the buffers (including upper-level management and clients) must understand that they are not slack. They must understand that the probability of on-time completion goes down if buffer sizes are reduced.

3. Historical buffer data can be very useful to analyze. The analysis can help identify areas where improvements are important, and can help reduce lead times. By knowing which tasks and which resources have caused buffer time to be used, we have useful places to look for process improvements. For example, if a specific resource is frequently behind, we can explore the causes, such as a tendency to underestimate task durations, or multitasking. We know that such improvements will help, because they have a direct impact on the buffers.

By knowing which resources frequently get ahead (or "unuse" buffer time), we can identify resources for which task durations tend to be overestimated. This not only helps identify additional capacity, but also allows overall lead times to be shrunk. However, be careful not to overinterpret the data: the existence of statistical fluctuations ensures that there will always be late and early tasks.

Historical buffer data can also be looked at more globally. If buffers are never used up completely, it's very possible that buffer sizes can be reduced, thereby reducing overall lead times.

Chapter 10

1. There are many ways of coming up with assumptions in this kind of conflict diagram. The "reference environment" approach directs you to think of an environment in which the arrow is not valid, then convert that environment into an assumption based on differences from your environment. We would read B to D as "In order to quote short lead times, we must avoid padding task times." A simple reference environment is the organization where salespeople don't care about what the lead times will really be, and give any quote they think will sell. The reference environment is "Sales doesn't care about reality"; in such an environment the arrow does not exist. The assumption is then "Quoted lead times must be closely related to actual lead times."

2. Any arrows can be attacked in Figure 10-1. For example, an assumption behind A to C is "Customers care about commitment dates." There are ways this assumption can be invalidated; for example, by lowering prices significantly. Such solutions may not always make sense.

3. One example is insurance. All the insured people pay some "average" amount. The insurance industry then collects enough money so that it can pay worst-case claims as needed. For most people this is much better than having to have funds available to cover the worst-case situation, which is like having padding everywhere.

    Mutual funds aggregate protection against fluctuations in investments. A single stock price can vary dramatically, resulting in huge potential swings in value. Aggregated stock prices are much more stable and therefore provide a greater degree of security.

Chapter 11

1. The schedule developed to this point should not be used as anything except a best-case, lower limit to the project duration. It contains no protection against uncertainty. The Critical Chain can be used to analyze how the project duration can be shrunk.

2. Parallel Critical Chain paths can exist. The problem is, the Critical Chain is not just the longest chain through the project network; it is also the set of tasks whose execution will be protected. Having parallel Critical Chain paths might make sense in situations where multiple paths should be monitored closely; otherwise, it will not.

   If two chains both seem to be critical, you will usually pick one; the other will be buffered so that the Critical Chain you have chosen is guarded.

3. When you don't know the capacity of a resource it makes no sense to resolve resource contention. This can happen with outside contractors. Even though the contractor has limited capacity, you will usually have no control over it. Sometimes it makes sense to build controls into the contract, such as penalties for being late, rewards for making the resources available, etc.

   Sometimes resources are effectively unlimited, because more can easily be obtained. For example, temporary workers are usually quick and inexpensive to hire; so for tasks that they perform (for example, copying, typing, or collating) it may not make sense to worry about resource contention. In some cases it may not even make sense to worry about task durations, when the tasks can be made very short by splitting them among several people.

Chapter 12

2. Buffers are essentially a worst-case estimate of how much longer the chain feeding the buffer might take, compared to the sum of the task durations. It is a kind of guess as to aggregate risk. It is dangerous to allow such guesses to be done strictly through an automated algorithm, because it is very difficult to quantify human understanding and intuition. The simple answer is "probably not."

3. Inserting the feeding buffers will cause Critical Chain gaps when there isn't room to push the chain feeding the buffer to the past. The effect of these gaps is to add protection to the Critical Chain. This protection could be considered a kind of informal buffer. Such informal buffers increase the chance of finishing the project on time, which suggests that the project buffer can be shortened.

4. Since the Critical Chain task durations do not include padding, any intermediate due dates that must be met require buffers to protect them, a kind of intermediate project buffer. Such buffers can be inserted. It is usually sufficient to specify a window of time during which the task is expected to be completed (like the "Emily Window" in Figure 8-3). This kind of window gives a realistic idea of when the milestone will be completed, without increasing the overall project duration.

Chapter 13

1. The objective will probably be something like "have an operational production facility." This is the objective that the company as a whole cares about. If you set "build the production facility" as an objective, it's likely that many tasks needed to get the facility operational (such as training, checking out equipment, etc.) will not be scheduled properly.

   Some major needs might be creating architectural drawings, constructing the facility, ordering equipment, installing equipment, and training workers. However, be careful: not all architectural drawings are required in order to start construction, and a completed facility is probably not needed in order to train workers.

3. Let's assume your discomfort arises because there's too much detail for you to have good global visibility over what's going on. This implies that a higher-level, aggregate picture is needed. If it's at all feasible, you should merge some of the detail to get a clearer picture. This merged picture doesn't have to be the picture that's given to individual workers. For a large program with many projects and thousands of tasks, it may make sense to think of the program as a whole as consisting of only the projects, with an overall program-wide Critical Chain. The individual projects can be scheduled separately with their own Critical Chains.

Chapter 14

2. Higher-level managers usually have a good sense of the goal of the company; otherwise they wouldn't have risen to their position. People lower in the hierarchy often don't have a good sense of what's really important for the company. Even if they do, they aren't rewarded for behavior that's globally positive. It's very common (for example) for construction workers to cut corners, engineers to overdesign, or computer programmers to add unnecessary features (and bugs), just through not understanding what's important.

    Awareness of the organization's goal helps people to work together to achieve improved performance. It is the first leverage point that should be addressed. Without a defined goal, improvement cannot be defined. Without widespread understanding of the goals, improvement may as well not be defined.

3. The value to the organization is zero.

Chapter 15

1. Throughput will look like a better leverage point than costs whenever (a) extra production capacity can be found, and (b) additional products can be sold (even below cost) without hurting existing markets.

2. A focus on quality by itself is unlikely to cause the right answer in the example. However, the focus on quality may be a result of the focus on throughput (see the section on Total Quality), which will give the better answer. If lifetime employment is considered a necessary condition for a company, the company must look to ways of increasing throughput (rather than cutting costs), which means throughput orientation is inevitable.

Chapter 16

1. Having to prepare everywhere for defense is like the modern manager having to look everywhere to cut costs. It is impossible really to do it effectively. There is no place where focusing helps significantly,

because costs can be cut everywhere. This is the Cost World point of view. Sun-Tzu's focus is on what is necessary to win the battle: gaining throughput. He was undoubtedly advocating a Throughput World approach.

The enemy should be prepared where he plans on doing battle; that is, where he plans on gaining throughput. If the enemy has no choice in (or knowledge of) where the battle will take place, he has a large disadvantage to overcome.

2. Every arrow has one or more assumptions. For example, the arrow between box 7 and box 4 contains (again) the assumption that the only way to deal with declining competitive position is to cut costs. We might read the arrow plus assumption, "**If** defense contractors are less and less competitive, **then** defense contractors try to cut costs, especially direct labor, **because** the only way to deal with declining competitive position is to cut costs."

3. "Best practice" is often similar to optimization. There is nothing wrong with exploring best practices in order to get new ideas. However, the sum of the individual best practices (the sum of the local optima) does not equal the overall best practice (the global optimum). This is often poorly understood.

    There is frequently a hope that taking a best practice and putting it in a different environment will produce benefits. This assumes both that the given best practice exists in isolation in the original environment, and that it will exist in isolation in the new environment. Neither assumption is true. You must therefore ask whether a given best practice will positively affect a leverage point. If not, it will do you no good.

4. Using Throughput/Operating Expense as a measurement gives an incentive both to increase throughput and to decrease operating expense. Unfortunately, using that ratio by itself can give a much stronger incentive to decrease operating expense. For example, halving operating expense would double the ratio, and often that is much quicker and easier than doubling throughput.

    This drawback can be gotten around by setting minimum targets for throughput. In other words, use of the productivity measurement seems to require at least one more measurement in order to make sure the proper incentives are given.

Chapter 17

1. In addition to the market leverage point, an organization will usually have at most one leverage point per independent chain. Each chain has at most one weakest link. Highly complex organizations tend to be very interconnected, and consequently there are few independent chains and very few weakest links. Simpler, less interconnected organizations tend to have more independent chains and therefore more weakest links. Paradoxically, the more complex an organization is, the fewer leverage points it tends to have, and therefore the fewer places where attention is needed in order to increase throughput significantly.

2. An organization that effectively incorporates the five-step process to improve continuously must be a Throughput World organization. We can expect numerous positive effects to flow from this, such as minimal layoffs, increased morale, and a tendency for policies to relate directly and logically to the organization's goal. For example, see Figure A-5.

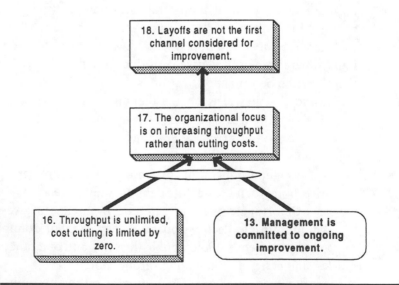

Figure A-5   The Throughput World and the Five Steps

Chapter 18

1. Adding the additional eight gizmos to the resource requirements gives Table A-1:

   Table A-1   Resource Requirements for Modified Figure 18-1

	Customer Service	Design Engineer	Computer Programmer	Technician	Total
Widget	12	12	40	24	88
Gizmo	48	48	40	48	184
Total	60	60	80	72	272

   With two programmers there is more than enough programming capacity even for the increased demand, and the new constraint is clearly the technician. The throughput per unit of constraint time for the technician is $40,000/3 weeks = $13,333/week for widgets and $30,000/3 weeks = $10,000/week for gizmos. So the company should make all 8 widgets, which leaves enough technician capacity for 8 more gizmos. This results in total throughput of ($40K × 8) + ($30K × 8) = $560K, the same as before. However, since operating expenses have increased, the net profit is now only $60K. This was a bad elevation decision. It would look like an even worse decision if you still believed gizmos were the most desirable product.

2. Dollars per unit of constraint time will cause you to take questionable actions if there are flawed underlying assumptions. One flawed assumption can arise if you don't pay attention to Step 5 of the five steps: "Before making any significant changes, **evaluate** whether the leverage point(s) will and should stay the same." If you sell so many profitable products that another resource becomes a constraint, your decisions based on the old (wrong) constraint will be flawed. For example, suppose the market in Figure 18-1 shifted so that there is demand for 16 gizmos and only 2 widgets. Where is the resource constraint? Which product is more desirable to produce?

   Another flawed assumption can occur if there are interactive constraints, constraint resources that feed one another. Interactive constraints are rare, but when they exist analysis according to throughput per unit of constraint time does not work.

3. According to throughput pricing, any amount for which you can sell free products above the truly variable costs will improve the bottom line. However, if you start selling products much more cheaply than your usual price and existing clients find out, they may demand a rebate. They will at least want the new price. Therefore, you must find ways of selling similar products for different prices without killing existing markets. There are many ways of doing this, such as changing product features, selling in different geographical markets, selling with different lead times, and so on.

4. A key factor missing from the traditional budget is some measure of throughput. The impossibility of estimating throughput precisely makes it impossible to budget precisely using throughput. Nevertheless, the traditional budget implicitly pretends something about throughput — for example, that it will not change. This is also a gross approximation. It is much better to admit that you don't know exactly what throughput will be and allow for the uncertainty than it is to pretend it doesn't exist.

Chapter 19

1. In this situation, there is no control of work coming into the organization. That means work-in-process continues to go up, because resource A will over the long run outperform resource C. Therefore, lead times continue to go up and competitive edge goes down. The workers who are idle the least will be those of resource A, because there is always new work to do. However, the entire organization (and everyone in it) loses as its ability to compete deteriorates.

Chapter 20

1. Here are some sample problems:
   - Task time estimates given by resource managers will tend to be worst-case elapsed times.
   - Shared resources will continue to multitask, requiring larger buffers.
   - Even for nonshared resources, resource managers may have trouble accepting the nonmultitasking paradigm.
   - There is little incentive for resources to work as quickly as possible when they have work.

None of these problems should make implementation of Critical Chain concepts impossible, but they can make it more difficult and reduce its effectiveness.

2. Plenty of capacity everywhere means that, most likely, interproject resource contention does not need to be resolved explicitly. This means it's less crucial to have a strategic resource schedule. Furthermore, in this situation the market is a very clear constraint and must be given highest priority; that implies an emphasis on the Critical Chain side of the scheduling. On the other hand, if the available capacity can be sold, it likely still makes sense to select a strategic resource. That way the mix of products to be sold will fit with the company's strategic plans, in anticipation of the time when there isn't plenty of capacity everywhere.

Chapter 21

Table A-2   Relating Schedule Components to Figure 19-2 Rules

Needed Schedule Components	Rules for Figure 19-2
Start times for tasks with no predecessors	(1) and (2): A may only start tasks when there are pennies in the "Project Starts" circle
The priority, if multiple tasks exist	N/A (first-come, first-served)
Who gets the work next	N/A (fixed by Figure 19-2)
Sometimes, approximately when your next job is coming and what it is	(3) B, C, and D should work on everything they have
Sometimes, how urgent a job is	N/A (first-come, first-served)
Task descriptions and requirements	N/A (always the same)

1. See Table A-2.
2. Here are some important concepts:
   - Buffers are necessary schedule components.
   - Buffers represent the uncertainty in a schedule.
   - It is inevitable that there will be people who aren't working on high-priority tasks.
   - Information about the global picture should be spread widely.
   - Information is power; relevant information can empower employees to work together more effectively.
   - Worry about completing projects on time, not tasks.

3. The "hot potato" must be passed as quickly as possible from one person to another. Material must keep flowing into the "assembly line" on time, or else the line stops. In a long "relay race" the baton must be passed as smoothly as possible, and directions on where to go next must be clear.

Chapter 22

1. This becomes very much like measuring whether people complete individual tasks in the time estimated. On the one hand, there seems to be an incentive to try to complete tasks as quickly as possible. On the other hand, there is an increased incentive to pad time estimates and to try and blame other people. The reality of statistical fluctuations must be accepted. Using average estimates of task durations, some tasks will be completed before and some after the average time.
2. This is an interesting measurement, in that it encourages the roadrunner mentality. However, it also encourages people to avoid accepting work and to declare tasks complete before they've been fully checked out. We can't imagine additional measurements that would make this one work.
3. Take the perspective of a customer, and consider what kinds of things they would look for in deciding whether they're delighted with your products. For example:
   • Lead times (how far ahead existing capacity is booked)
   • Reliability (defect rates, rework rates)
   • Performance relative to expectations (product returns, complaints)
      While some of this terminology looks more appropriate for manufacturing, these measurements can be applied to many different types of project environments.

Chapter 23

1. A SWAT team can require that some people have more than one task active at the same time. However, the priorities should always be very clear, so the worst effects of multitasking (everything is completed late) are not present.

2. It can be bad to split tasks apart if it takes significant time to prepare additional people to work on the subtasks. It might take months for a new person to become productive in a project; this can be considered "setup" time. Task splitting can also be bad if there must be significant communication between the people working on the subtasks, as discussed in Chapter 4.

3. There are a couple of key assumptions behind this arrow: that happier workers are more productive, and that the productivity of all workers contributes to company profits. The first assumption can be attacked by assuming workers are mindless automatons; the second by trying to keep only key workers happy. There are usually drawbacks to both of these approaches.

Chapter 24

1. It is hard to judge whether officers are held in more esteem than civilians. But assuming we are talking about solutions of similar complexity, officers will generally spend only a few years suffering from any given problem. The NIH law indicates that officers would on average have the least resistance to new solutions.

2. Probably before you buy the elephant, certainly before you kill it.

3. If someone wishes to make work expand to fill some available time, one approach is to "polish" or add new features to existing products. This is common among engineers and computer programmers. However, there are a number of unfortunate ramifications of this behavior, such as more errors to fix, more documentation to be done, more customer training, and less free time for the worker. If the workers understand "good enough," they will be less likely to indulge themselves in this way.

Appendix B: Glossary

Buffer   A buffer is time put into the schedule systematically in order to protect against unanticipated delays, and in order to allow for early starts. Buffers are **not slack**; they are essential parts of the schedule. There are four types of buffers: project buffers, feeding buffers, resource buffers, and strategic resource buffers.

Conflict Diagram   A logical tool that helps to solve conflicts in a win–win manner. See also *Evaporating Cloud*.

Constraint   A constraint, according to Theory of Constraints, is anything that limits your performance relative to your goal. For a for-profit organization, a constraint is usually anything which prevents the organization from generating better bottom-line results. There are many types of constraints, including policies, people, and equipment. In this book we usually refer to constraints as "leverage points" when they can be exploited to leverage a better bottom line.

Note that this meaning is very different from another meaning of "constraint" that is in common use. Sometimes a constraint can be taken to mean any restriction placed on a schedule. For example, a "task timing constraint" is a restriction placed on task start and/or end times, such as "must start before" or "must finish on." Usually such constraints do not limit performance relative to the goal.

Core Problem   A problem which causes many undesirable effects to exist: see also Current Reality Tree.

Cost World   The Cost World is an in-the-box mentality that diverts attention to the many things that can lower costs, and away from the few things (leverage points) that will improve throughput.

Critical Chain   The Critical Chain is that set of tasks which determines the overall duration of a project. Usually it requires taking resource capacity into account. It is typically regarded as the constraint or leverage point of a project.

Critical Path Method   The Critical Path Method (CPM) is a technique for analyzing projects by determining the longest sequence of tasks (or the sequence of tasks with the least slack) through a project network. It is typically used as a means of determining which tasks should be concentrated on.

Current Reality Tree   This is a useful tool both in exploring and communicating the cause-and-effect relationships of a given subject, and in identifying a small number of core problems that cause a much larger number of undesirable effects.

Dependent events   Dependent events are events that depend directly on one another, either due to a predecessor–successor relationship (one must come first), or a resource relationship (they depend on the same resources). Dependent events are the source of interconnections in and between projects. Highly interconnected environments have few leverage points.

Elevate   This is the fourth step of the five-step focusing process. Elevation of a leverage point means increasing operating expense to obtain more capacity. Overtime, hiring new workers, and buying new equipment are examples of elevation.

Evaluate   This is the modified fifth step of the five-step focusing process. The likely effect of actions should be evaluated in advance to determine their impact on leverage points.

Evaporating Cloud   A logical tool that helps to solve conflicts in a win–win manner. First the conflict is diagrammed, by verbalizing the conflict (prerequisites), the reasons for the conflict (requirements), and the overall objective that causes both requirements to be needed. Assumptions are surfaced by examining the connections between objectives, requirements, and prerequisites; those assumptions are then challenged to develop breakthrough solutions. Also known as a Conflict Diagram.

Exploit   Step 2 of the five-step process. A leverage point is exploited by squeezing more out of it.

Feeding buffer   A feeding buffer is placed at each point where a non-Critical Chain task joins the Critical Chain. This buffer protects the Critical Chain from disruptions on tasks feeding it, and allows for early task starts.

Five-step process   The five-step process is a means of improving a system over time by identifying, focusing on, and improving leverage points. The steps are identify, exploit, subordinate, elevate, and go back to step 1 (don't let inertia become a constraint). The steps can be revised to provide more stability; the revised five steps are select, exploit, subordinate, elevate, and evaluate.

Free Product   Products that do not depend on the leverage point and therefore are not constrained by existing capacity are known as "free products." Any amount they're sold for above materials costs goes straight to the bottom line.

Future Reality Tree   This is a tool that helps examine and communicate the expected effects of proposed changes or "injections." It is similar to the Current Reality Tree, except it deals with future events and effects.

Gating tasks   Tasks that have no predecessors are known as "gating" tasks because they are the gates that control the flow of work into the system.

Goal   The reason a system exists. The owner of a system defines its goal. The goal of most public companies is to make money now and in the future.

Good enough   Better than which, who cares?

Identify   This is Step 1 of the original five-step focusing process. You must identify leverage points before you can use them to improve.

Injection   An injection is a change that is "injected" into the current reality, in order (one hopes) to create a desirable future reality. In order to state and examine an injection, it is not necessary to prove that it is possible.

Intermediate due dates   Dates on which certain project tasks or deliverables must be completed, not including the project due date.

Investment   This is money an organization invests in items that retain salable value, such as buildings, equipment, and materials. It can be converted to operating expense through depreciation, or throughput through appreciation.

Lead time   This is the time by which a product start must "lead" the customer commitment date in order for the product to be delivered on time.

Leverage point   A leverage point is an area where a small change can have a big positive impact on the bottom line. Leverage points can also be called "constraints."

Market segmentation   In TOC terms, market segmentation is selling the same product for different prices, usually in different markets, without hurting existing markets.

Multitasking   Multitasking is the practice of giving people more than one task to do at the same time, without having a clear and consistent priority among the tasks. Among other things, it usually results in people taking a long time to complete every task, which results in longer-than-necessary project durations.

Operating expense   Operating expense, one of the three fundamental measures, is the rate at which money must be put into the system to keep it generating throughput.

Pareto Principle   The Pareto Principle or 80/20 rule indicates (when talking about improvement) that 80% of the impact can be achieved through 20% of the actions. In environments with dependent events and statistical fluctuations, it's more likely that 95% of the impact can be achieved through 1% of the actions. This principle implies the existence of leverage points.

Parkinson's Law   "Work expands so as to fill the time available for its completion."*

Precedence dependency   A precedence or path dependency between two tasks is the requirement that one task (the "predecessor") must be completed before the next (the "successor") can be worked on.

Prerequisite Tree   This is a tool that connects obstacles and needs with tasks (sometimes called "intermediate objectives"). It is useful both to put tasks in time sequence and in explaining why certain tasks are needed.

Process batch   This is a single, conceptually discrete piece of work that a person or work center works on. See also *Transfer batch.*

Project   Every book gives its own definition of project, which means we can't produce one that is entirely satisfactory. Intuition should be sufficient. Projects usually imply an objective and a set of tasks that must be coordinated to achieve the objective.

Project buffer   A project buffer is placed after the final task of a project in order to protect the project completion date from delays, especially delays along the Critical Chain. This is the most important buffer to place and to monitor.

Protective capacity   This is extra capacity needed on nonstrategic resources in order that they will not become constraints.

Resource   A resource is anything that is required to perform one or more tasks. Resources can be people, equipment, and buildings.

---

* Parkinson, C. N., *Parkinson's Law*, The Riverside Press, Cambridge, 1957, 2.

Resource buffer   The resource buffer is provided before resources start work on Critical Chain tasks to make sure that resources will be available and the Critical Chain tasks can start on time or (if possible) early. There are two ways the resource buffer can be implemented. It can be implemented as a wake-up call, which means the resource (or resource manager) is informed at regular intervals before the resource is needed for the Critical Chain task. This allows it to be prepared to start immediately. Alternatively, space (idle time) can be created on the resource for the buffer, in order to ensure the resource's availability for the Critical Chain task. This space could be considered a kind of protective capacity.

Resource dependency   A resource dependency between a task and a resource is the requirement that a resource be assigned to the task in order that the task can be worked on.

Select   In the revised five-step process, the first step is "Select the leverage point(s)." Decide strategically where they should be, and take actions — when it makes sense strategically — to keep them there.

Slack   Slack is free time available to move a task later, synonymous with "float." In this book we use the terms "past slack" and "slack to the past" to indicate free time available to move a task earlier.

Splitting and overlapping   Many times tasks can be completed more quickly if they are split across several resources and worked on simultaneously. For example, it might take half the time to paint a house if two people, rather than one, are doing the work. Frequently opportunities to split and overlap tasks are missed.

Statistical fluctuations   Unanticipated disruptions, or Murphy's Law: if anything can go wrong, it will. No organization is completely free of statistical fluctuations.

Strategic resource buffer   The strategic resource buffer is placed before tasks performed by the strategic resource, in order to protect the strategic resource from disruptions on nonstrategic resources. This protection helps maximize throughput for the organization as a whole. It is only intended for use in multiple-project environments. This is also known as a "bottleneck buffer."

Subordinate   This is Step 3 of the five-step process: subordinate everything else to the decisions made in the identify/select and exploit steps. Ultimately it means that everyone should do their best to support critical resources and the Critical Chain; that is, to support the organization's goals.

Task   Any element of work needed in order to complete a project. Sometimes the terms "task" and "activity" are used interchangeably; sometimes activities are considered to be made up of a number of tasks.

Theory of Constraints   Theory of Constraints is a management philosophy first developed by Dr. Eliyahu Goldratt. It provides tools and concepts that can help make people and organizations more productive according to their goals.

Throughput   Throughput, strategically the most important of the three fundamental measurements, is the rate at which a system generates money (through sales). "Through sales" is specified because a half-completed project should not count as having generated half the money.

Throughput lever   A throughput lever is a leverage point that helps to produce more throughput. Most of the powerful leverage points are throughput levers.

Throughput pricing   Throughput pricing uses time required on a constraint (or leverage point) as a means of assessing the "cost" of a product. This is done by looking at the ratio throughput per constraint unit, where the constraint units are expressed in time. Assuming no other capacity problems arise, products generating higher throughput per constraint unit will produce better bottom-line results.

Throughput World   The Throughput World is a mindset that is oriented toward producing more throughput. This mindset requires looking outside the box of the current system. It assumes that leverage points exist.

Transfer batch   This is a chunk of work that is transferred from one person or work center to the next. The transfer batch size does not have to be equal to the process batch size.

Work-in-process   Work-in-process (WIP) is work sitting in a system waiting to be finished. WIP usually has a significant impact on lead times, because typically the existing WIP must be finished before new work can be finished. In a project environment, WIP could be information (stored on paper, computer disk, etc.) as well as physical inventory.

Appendix C:
ProChain™ Project Scheduling

We have developed a software scheduling tool called "ProChain™ Project Scheduling"* which combines with Microsoft® Project to enable easier implementation of Critical Chain scheduling techniques. Elsewhere in the book we describe some of the practical ramifications and benefits of buffer management, critical chain scheduling, and so on. In this Appendix we describe two of the lessons we have learned from the implementation of this software. The first is nature and magnitude of the shift in thinking required in moving to Critical Chain scheduling. This shift can be described as moving from pushing work into the front of a system, to pulling work through the system. This change is especially difficult to make when the traditional paradigm is supported by existing software systems. The second lesson is the value of analyzing the project network using Critical Chain thinking.

C-1   Push-vs.-Pull

As project managers scheduling projects, most of us have grown up in a "push" environment. This is an environment in which work is pushed into the system as quickly as possible. Unless a task requires significant investment, it will be started (but probably not completed) as early as possible.

We have all experienced problems with this approach. First, there is always uncertainty regarding how long tasks will take. There are things you just can't know in advance. Second, resource contention is typically not resolved formally,

---

* For further information, consult the web page "http://www.prochain.com".

which means the schedules can't be realistic. To resolve resource contention the project managers and/or the resource managers must make immediate prioritization decisions, although it's not always clear what the priorities should be. The net effect of not resolving the resource contention, even in the best case where multitasking doesn't result, is more uncertainty. We can't predict when things are going to happen. We typically deal with that uncertainty by adding "safety" to task times, and making them into worst-case times.

Some characteristics of the standard push system schedule are:

- Jobs are started as early as possible.
- Resource contention is not resolved as part of the initial schedule.
- There is no good basis for prioritizing tasks for specific resources.
- Schedules quickly become wrong as the situation changes.
- Each individual task completion date is considered important.

Organizations responds to these characteristics in various ways:

- By default, tasks are scheduled as early as possible.
- Task durations are have safety times added.
- As a result of padding task durations, workers must multitask. This, in turn, makes virtually impossible to resolve resource contention.
- Schedules are revised frequently, to account for unexpected circumstances.
- Many pieces of the project plan are not worked out, because the schedule will change anyway.
- Resource contention is sometimes resolved by putting in artificial dependencies between tasks, or artificial task constraints such as "start-no-earlier-than".
- Project managers must be "squeaky wheels" in order to get resources they believe they need.
- Projects are frequently late, over budget and under performance.

In the "pull" system, work is released into the system as late as is practical, taking into account the ability of the entire system to process it. Only work that needs to be done is done, and it is done only when needed. It's useful to reiterate some important characteristics of our pull system:

- Average time estimates are used.
- Scheduling is done backwards from the date a project is needed, rather than forwards from today.
- Buffers are used to protect the entire project and the Critical Chain, rather than individual tasks.
- Schedules don't need to change often, because they are protected by buffers.
- Workers are encouraged to work as quickly as possible; however, they are not penalized if they are late.

Most existing computerized scheduling systems are designed to implement the "push" system. Even though implementing the Critical Chain approach may result in only small changes to the underlying data management system, it implies major changes in how the system should be used. Some of the significant differences between push-system schedules and pull-system schedules are shown in Table C-1. These differences need to be well understood. Any of them can make Critical Chain schedules look peculiar to those familiar with traditional Critical Path schedules.

Table C-1   Push vs. Pull

Push System	Pull System
Individual tasks are protected by adding safety times.	Tasks are scheduled without safety times; the project is protected by adding buffers.
Tasks are expected to be placed at their early start times.	Tasks are placed as late as possible, while leaving room for buffers.
Resource contention is ignored, if not explicitly modeled through task-to-task links.	Resource contention is taken into account.
Worst-case task durations are used.	Average task durations are used.
Task-to-task linkages incorporate scheduling decisions, by reflecting not just how the tasks must be sequenced, but how the project manger wants them to be sequenced.	Task linkages reflect what is possible and do not include scheduling decisions.
Multitasking, or working on several tasks at once, is normal.	Multitasking is not done; instead, the schedule indicates clear priorities between tasks.

Table C-1   Push vs. Pull (continued)

Push System	Pull System
Task constraints are sometimes added in order to override the standard early start logic.	Task constraints reflect only hard scheduling requirements.
Unexpected events are taken into account by creating new schedules.	The impact of unexpected events is monitored by looking at the buffers.
Critical path tasks, which are tasks having no future slack or "float", are highlighted in the schedule as being important.	Critical chain tasks, which are tasks having no past slack, are highlighted in the schedule as being important.
Start tasks at the scheduled start time; finish at the scheduled finish time.	If a task has predecessors, start as early as the work is ready. If not, start no earlier than the scheduled start time. Finish as quickly as possible.
Task priorities are according to what seems important today, and may change frequently.	Task priorities are according to the Critical Chain and the scheduled start times.

C-2   Preparation of a Project Network

A "project network" is a formal plan of the tasks, resources and linkages needed to achieve the objective(s) of the project. Such a plan can be created using the prerequisite tree approach described in Chapter 13.

It's not uncommon for projects to be performed, from start to finish, without ever creating a project network. Even when there is a project network, it usually includes some bad modeling decisions which will push out the project due dates; decisions like unnecessary task links, unwillingness to overlap work and inflated task times.

The use of Critical Chain project scheduling, especially using a computerized scheduling tool, makes it possible more easily to analyze means of improving the project network. For example, you can:

- Identify the extent to which task time estimates are inflated: compare the desired project completion date with the scheduled, buffered completion date. Look at the individual Critical Chain tasks. Analyze the critical tasks, especially the long ones, to see how much hands-on time will really be spent on the work.

- Look for tasks that can be split and overlapped to reduce the project duration, especially long tasks on the Critical Chain.
- Look for task linkages that push the project out, but which aren't "real": links between Critical Chain tasks.
- Identify resource contention problems which will cause the project to be moved out: look for places where the Critical Chain hops from one path to another. These discontinuities may signal places where the project duration can be shortened by adding resources.

There are also some traps that a computerized tool may help you to fall into:

- It's easy to go into great detail, because much of the drudgery is done by the computer. Resist this urge; it can cause you to spend much time with little benefit. There is always more detail.
- Frequently people believe things because a computer says they're so. A computerized plan is only as good as the data input, which always contains questionable assumptions. Check the results at every phase of the scheduling process.
- If the resulting plan is not satisfactory (the first clue is usually that it's too long), it is common to blame the software or the methodology rather than the assumptions behind the project network. That is counter-productive and leads inevitably to failure: the methodology doesn't work, we make "adjustments," it works more poorly, we make more "adjustments," and so on.

Keeping a global perspective and a "good enough" philosophy will help you avoid these traps.

Index